Highwall Mining

T0200533

Highwall Mining

Applicability, Design & Safety

John Loui Porathur
CSIR-CIMFR Regional Centre, Nagpur, India

Pijush Pal Roy
CSIR-CIMFR, Dhanbad, India

Baotang Shen
CSIRO, Brisbane, Australia

Shivakumar Karekal
University of Wollongong, NSW, Australia

CRC Press
Taylor & Francis Group
Boca Raton London New York

CRC Press is an imprint of the
Taylor & Francis Group, an **informa** business

A BALKEMA BOOK

Cover photo: ADDCAR continuous highwall miner in operation at Ramagundam Opencast Project-II of M/s Singareni Collieries Company Ltd. (SCCL). Courtesy of Highwall Mining Group of CSIR-CIMFR.

CRC Press
Taylor & Francis Group
6000 Broken Sound Parkway NW, Suite 300
Boca Raton, FL 33487-2742

First issued in paperback 2019

© 2017 by Taylor & Francis Group, LLC
CRC Press is an imprint of Taylor & Francis Group, an Informa business

Typeset by MPS Limited, Chennai, India

No claim to original U.S. Government works

ISBN-13: 978-1-138-04690-0 (hbk)
ISBN-13: 978-0-367-88932-6 (pbk)

Library of Congress Cataloging-in-Publication Data

Visit the Taylor & Francis Web site at
http://www.taylorandfrancis.com

and the CRC Press Web site at
http://www.crcpress.com

Table of contents

List of Abbreviations

AB	Quarry AB, West Bokaro, Tata Steel Ltd. (India)
ACARP	Australian Coal Association Research Programme
ADDCAR	ADDCAR Systems LLC (Kentucky, USA)
AISRF	Australia-India Strategic Research Fund
AMI	auger mining index
AMT	Advanced Mining Technologies Pvt. Ltd. (Hyderabad, India)
ANZECC	Australian and New Zealand Environment Conservation Council
ASTM	American Society for Testing and Materials
BHP	BHP Australia Coal Pty. Ltd. (Brisbane, Australia)
BM	barren measure
CHM	continuous highwall mining
CIMFR	Central Institute of Mining and Fuel Research (India)
CMRI	Central Mining Research Institute (India)
CMRR	coal mine roof rating
CRFM	coal roof failure model
CSIRO	Commonwealth Scientific and Industrial Research Organisation
DFN	discrete fracture network
DGMS	Directorate General of Mines Safety (India)
ECM	equivalent continuum medium
EGAT	Electricity Generating Authority of Thailand
FFT	fast Fourier transform
FISH	built-in scripting language in FLAC
FLAC	numerical modelling software (Fast Lagrangian Analysis of Continua) (ITASCA)
FoS	factor of safety
HMI	highwall mining index
HMR	highwall mining reserve
HWM	highwall mining
IEL	Indian Explosives Ltd. (Gomia, India)
ISEE	International Society of Explosives Engineers
ISRM	International Society for Rock Mechanics
ITASCA	ITASCA Consulting Group Inc. (Minneapolis, USA)
ITB	Institute of Technology, Bandung (Indonesia)
KTK	Kakatiya Khani (Warangal, India)

LIPI	Lembaga Ilmu Pengetahuan Indonesia (Indonesian Institute of Sciences)
LSFM	laminated span failure model
MOCP	Medapalli Opencast Project (Ramagundam, India)
MSHA	Mines Safety and Health Administration (USA)
NEW	non-effective width
NIRM	National Institute of Rock Mechanics (India)
NX	standard core size of diameter 54.7 mm
OCP	Opencast Project
PDF	probability distribution functions
PPV	peak particle velocity
QWQ	Queensland Water Quality
RCC	reinforced concrete column
RMCHM	rigid modular continuous highwall miners
RMR	rock mass rating
RMS	root mean square
SCCL	Singareni Collieries Company Ltd. (Kothagudem, India)
SEB	Quarry South-Eastern Block, West Bokaro, Tata Steel Ltd. (India)
SECL	South Eastern Coalfields Ltd. (Bilaspur, India)
SF	safety factor
SG	specific gravity
SGLP	synthetic groundwater leaching protocol
SOM	self-organising map
TCLP	toxicity characteristics leaching procedure
TSL	Tata Steel Ltd. (Mumbai, India)
UCS	unconfined compressive strength
UDEC	numerical modelling software (Universal Distinct Element Code) (ITASCA)

Preface

Several open-pit coal mines, around the globe, are reaching their pit limits. Existence of surface dwellings in many places limits the expansion of currently running open-pit mines. Also, in numerous cases, the overburden becomes so high that coal extraction becomes uneconomical.

Continuous highwall mining (CHM) is a relatively new technology which can extend the life of opencast mines without disturbing the surface dwellings, while maintaining economy and productivity. It is a remotely operated coal mining technology which consists of the extraction of coal from a series of parallel entries driven into the coal seam from the face of the highwall. These entries are unmanned, unsupported and unventilated. This technology uses highwall machines where a cutter is placed on the top of a continuous miner and taken through a conveyor inside the seam, which can be almost 500 m deep inside. In the present day, penetrations up to 500 m have been consistently achieved with highwall mining systems, in contrast to auger mining wherein penetrations are limited to 100–150 m.

CHM technology has been in use in the USA and Australia using Addcar, Cat/HW 300 and Bucyrus Highwall Mining System for the past couple of decades, and was recently introduced in several other countries including India, where the domestic energy requirement is met by coal combustion to the extent of about 60%. To meet the growing energy requirements, India is keen to develop all available technologies to help achieve the ambitious targets. In contrast to other sites worldwide where CHM has been introduced, Indian coal mining has complex geo-mining conditions with multiple, contiguous and thick seams.

The first continuous highwall mining system in Asia started on 10th December 2010 at Opencast Project-II (OCP-II), Ramagundam Area-III (RG-III) of Singareni Collieries Company Ltd. (SCCL), in India. This was followed by Sharda OCP of Coal India Ltd. and a few other mines. Up until the present, more than 3.5 million tonnes of coal costing over ₹550 crore (US$ 81 million) have been recovered in India. The outsourcing charges vary between ₹1,000 to ₹1,200 (≈US$ 15–18) per tonne of coal production depending on the site conditions, which is considered to be very economical compared to other underground methods of winning coal.

Due to there being no comprehensive technical book on highwall mining in the world, the authors felt it necessary to publish an international standard book on highwall mining covering theory and practice, coupled with case examples and design methodologies. The book contains eight extensive chapters covering the description of highwall mining, world scenarios, economic potential, methods of coal extraction and design methodology, including empirical web pillar design, numerical modelling for stress analysis, safety factor for web pillars, panel and barrier design, small- and large-scale numerical modelling, multiple seam interaction and design, coal web pillar strength, equivalent width concept, new web pillar strength formula, effect of weak bands in coal seam, slope stability, safety and ground monitoring, laboratory test results, case examples, hazards and regulatory requirements, and norms and guidelines for practice. It also summarises the results of research carried out on the subject by the CSIR-Central Institute of Mining and Fuel Research (CSIR-CIMFR), India, and the Commonwealth Scientific and Industrial Research Organisation (CSIRO), Australia.

Many people deserve appreciation for the structural formation and development of this emerging technology. Primary recognition goes to the other scientists of the Highwall Mining Research Group of CSIR-CIMFR, namely, Dr Jagdish Chandra Jhanwar, Senior Principal Scientist; Dr Amar Prakash, Principal Scientist; Dr Chhangte Sawmliana, Principal Scientist; Mr John Buragohain, Senior Scientist, and Dr (Mrs) Chandrani Prasad Verma, Principal Scientist. A few research outcomes of the doctoral thesis of Dr (Mrs) Chandrani Prasad Verma on highwall mining under the guidance of one of the authors have been thankfully cited in a few places. Munificent help, support, guidance and coordination of Dr Amalendu Sinha, Former Director, CSIR-CIMFR; Dr Hua Guo, Research Director – Coal Mining, CSIRO; Mr N.V. Subba Rao, Director, AMT, Hyderabad and Mr D.L.R. Prasad, Former Director (Planning & Projects), SCCL are gratefully acknowledged. Their unstinting guidance helped the research team to obtain the DST-sponsored Indo-Australia Joint Project on "Highwall mining design and development of norms for Indian conditions", which became a strong basis for writing this book for the benefit of the coal mining industry at large. Thanks also go to the colleagues and friends who were directly or indirectly involved in this work, namely, the late Dr P.R. Sheorey, Former Chief Scientist of CSIR-CIMFR; Mr Nandeep Nadella, Vice President – Operations, AMT; Mr Hari Prasad Pidikiti, Director, DCSL; Dr Angad Khuswaha, Chief Scientist, CSIR-CIMFR; Dr V.K. Singh, Chief Scientist, CSIR-CIMFR; Dr R.V.K. Singh, Chief Scientist, CSIR-CIMFR; Dr N. Sahay, Chief Scientist, CSIR-CIMFR; Dr Marc Elmouttie, CSIRO Energy, Australia; Mr Sungsoon Mo, University of New South Wales; Mr Zhiting Han, University of Queensland; Mr B.N. Pan of M/s Cuprum Bagrodia Ltd. and other colleagues of CSIR-CIMFR who have extended all possible help and guidance in the preparation of the manuscript. Dr Chhangte Sawmliana, Principal Scientist, CSIR-CIMFR and Dr Amar Prakash, Principal Scientist, CSIR-CIMFR deserve appreciation for their enumerable help in tracing few diagrams and contributing some scientific inputs of this book. The valued research contributions made by several researchers worldwide have also enriched the contents of the book. Some parts of the book are reproduced from AISRF Final Report: Agreement ST050173 carried out by the authors.

Dr Pradeep Kumar Singh, Director, CSIR-CIMFR; Mr Rahul Guha, Former Director General of Mines Safety, DGMS; Mr S.J. Sibal, Former Director General of Mines Safety and Mr S.I. Hussain, Deputy Director General of Mines Safety, DGMS were the sources of inspiration of this total task. Without their encouragement it would not have been possible to write this book.

We are especially indebted to Dr Marc Elmouttie, Research Team Leader, CSIRO Energy, Brisbane, Australia for his kind consent to present his work on Sirovision™

and Discrete Fracture Network (DFN) Modelling as part of the AISRF Final Report: Agreement ST050173 in Chapter 5 for global readers.

The objectives of the book remain in understanding the complex, multiple seam scenarios for highwall mining with maximum coal recovery from any given site with better economics, thereby helping mining companies to frame norms to avoid hazards and minimise instability issues. We are confident that this book—which is the first of its kind—will prove to be most helpful as a text/reference book for all those concerned, and if this happens to be the case, we will consider our paramount efforts to be for a noble cause in the understanding of highwall mining operations for the safe, efficient and economic extraction of locked-up coal in many countries around the globe.

John Loui Porathur, Pijush Pal Roy
Baotang Shen, Shivakumar Karekal

Foreword

When an opencast mine is excavated, a highwall remains after excavation, which, in most cases, is either abandoned or covered up. This eventually motivated the management of coal mining companies to devise some means of extracting coal from an existing highwall, which would otherwise be lost. Highwall mining is a method that originated from auger mining. It is a new, remotely operated coal mining technology which can extend the life of opencast mines without disturbing the surface structures, while simultaneously maintaining economy and productivity. The method consists of the extraction of coal from a series of parallel entries driven into the coal seam from the face of the highwall. The operation is carried out by remote control at the surface where an operator sitting in a cabin uses on-board cameras to monitor and control the progress of the continuous miner.

Over the last few years, a joint research team involving scientists of the CSIR-Central Institute of Mining and Fuel Research (CSIR-CIMFR), India and the Commonwealth Scientific and Industrial Research Organisation (CSIRO), Australia, has put all their efforts into carrying out a joint R&D project funded by the Australia-India Strategic Research Fund (AISRF), which resulted in the formation of comprehensive guidelines for highwall mining. CSIR-CIMFR has also carried out substantial work in the design and field implementation of highwall mining technology at sites in Singareni Collieries Company Limited, South Eastern Coalfields Limited and Tata Steel Limited. The knowledge and experience gained through this joint research project has already appeared in various quality publications. In addition, the Directorate General of Mines Safety (DGMS), India, issued Technical Circular No. 06 of 2013 (dated 22.08.2013), on norms and guidelines for highwall mining based on the studies conducted in India.

Taking stock of the need of the hour, the authors have documented their vast experience and knowledge in this new technological domain in the form of an invaluable book which contains eight extensive chapters covering fundamental, analytical and practical aspects of highwall mining operations under varying geo-mining conditions. Along with many interesting and innovative case studies, it explains, quite lucidly, some promising areas with future prospects and scope for improvement. The book will undoubtedly bridge the knowledge gap in respect of safe, efficient and cost-effective contemporary technology on highwall mining.

It is also noteworthy to mention here that over 3.5 million tonnes of coal costing approximately US$ 81 million has already been extracted using this technology from four sites in India successfully, safely and cleanly through the guidance of CSIR-CIMFR. Several coal mining companies are now showing interest in implementing this technology due to its high standard of safety, ease of extraction and good profitability.

It is my strong conviction that more advancements will be made in this technology in the near future, which will transform the clean coal extraction programme successfully while establishing it as one of the mainstream coal mining methods.

I am confident that due to its broad coverage on design simulations and discussions on various issues starting from production, productivity, economics, operational hurdles and safety, this book will be treated as a unique storehouse of knowledge in respect of highwall mining technology. The authors, without any iota of doubt, deserve high kudos for such exemplary contributions in this emerging area.

Kolkata

Sutirtha Bhattacharya
Chairman, Coal India Limited

Foreword

Safety, productivity and environmental responsibility are the three predominant considerations for modern mining operations worldwide. Highwall mining is an emerging coal mining method that offers distinctive advantages in the following three areas: it is a remotely controlled system that removes people from hazardous areas; it is a low-cost mining method that does not require large capital investments or a long lead time; and it is a low-impact method on the overburdened rock and ground surface compared with traditional open-cut or longwall mining methods.

The book "Highwall Mining: Applicability, design and safety" is a very timely reference tool for the coal mining industry. Although highwall mining has been practised for over 20 years in the US and Australia—and more recently in India—there hasn't, to date, been any systematic reference book published to help the design and assessment of highwall mining systems. This book, for the first time, provides readers with systematic knowledge and methodologies for highwall mining; from the initial feasibility assessment and layout design, to operational stability monitoring. It covers extensive research results and the operational experiences of more than two decades worldwide, and is invaluable to both the industry and research community alike.

This book is a result of a recent Australia-India Strategic Research Collaboration Project sponsored by both the Australian and Indian governments, and conducted jointly by CSIRO and CSIR-CIMFR. It is authored by four excellent researchers from both organisations who have been working on highwall mining for many years. The rich experience and previous research results of highwall mining in Australia since the 1990s have been further extended for Indian geological and mining conditions, and new design guidelines have also been developed by the authors for applications more broadly beyond Australia and India. All these latest developments have been thoroughly presented in this book. It is my belief that the book will offer the most advanced knowledge in highwall mining for a long period to come.

Brisbane

Dr Hua Guo
Research Director – Coal Mining
CSIRO Energy

About the authors

Dr John Loui Porathur is a Principal Scientist at the Regional Center for Rock Excavation and Rock Mechanics at Nagpur, India under CSIR-Central Institute of Mining & Fuel Research, Dhanbad, India. He has both a Bachelor of Technology (Hons) and PhD degrees in Mining Engineering from the Indian Institute of Technology, Kharagpur, India. Since 1997 he has been working as a scientist in the CSIR-Central Institute of Mining and Fuel Research (CSIR-CIMFR). His major areas of research include rock mechanics, mining methods, numerical modelling, and subsidence engineering, etc. Dr Porathur has been the recipient of many prestigious national awards, including the CSIR-Young Scientist Award in 2003, and the Team Leader for the CSIR Technology Award in 2011; the latter for his valuable contributions to developing new technologies in the field of mining engineering. He has published and presented over 65 research papers in various international and national scientific journals and conferences (e-mail: johnlouip@gmail.com).

Dr Pijush Pal Roy is an Outstanding Scientist and Head of the Research Group of Rock Excavation Engineering Division of the CSIR-Central Institute of Mining & Fuel Research, Dhanbad and former Acting Director at the CSIR-Central Mechanical Engineering Research Institute, Durgapur. Before joining CSIR-CIMFR, he was in the teaching profession for nearly five years, and has 31 years of R&D experience in CSIR-CIMFR. Dr Pal Roy obtained his PhD degree in Applied Sciences from the Indian Institute of Technology (Indian School of Mines), Dhanbad in 1984. He is the author of three books in rock blasting published by Oxford IBH Publishing Co. Pvt. Ltd. and Taylor & Francis Group (CRC Press), as well as two guidelines and 124 national and international publications. He is the recipient of the CSIR-Young Scientist Award (1989); the CSIR-Golden Jubilee First CMRI-Whittaker Award (1993); the National Mineral Award (2005); the CSIR Technology Award (2011); and the Hindustan Zinc Medal of the Institution of Engineers (1997 & 2012). His research areas include rock blasting technology, highwall mining, mathematical modelling and the environmental impacts of blasting. He has worked in more than 320

mines and quarries, some 15 hydroelectric projects as part of sponsored and consultancy projects, and many S&T Projects sponsored by the Government of India (e-mail:
ppalroy@yahoo.com).

Dr Baotang Shen is a Senior Principal Research Scientist and
group leader for Mine Safety and Environment at CSIRO
Energy. Dr Shen obtained his first degree in mining engineering in 1985 in China. He obtained his PhD in rock
mechanics in Sweden in 1993, and then worked with the
Norwegian Geotechnical Institute (NGI) as a research fellow
in 1994–95. Since 1995 he has been with CSIRO where he
focused on issues related to highwall mining geo-mechanics,
mining engineering and the mining environment. Dr Shen
specialises in mining engineering, monitoring instrumenta-
tions, numerical modelling, backfill technologies and rock
fracture mechanics. Since 2004, he has been leading CSIRO's
research team on mine subsidence control and mine remedi-
ation, and developed the non-cohesive backfill technology for old mine remediation.
The team has completed several major projects funded by the Australian Coal Association Research Program (ACARP), BHP Billiton Illawarra Coal and the Queensland
state government, on subsidence control and remediation technologies for a number of
mines in Australia. Dr Shen is an influential expert in fracture mechanics modelling. He
is also the leading author of the recently published book "Modelling Rock Fracturing
Processes" (e-mail: baotang.shen@csiro.au).

Dr Shivakumar Karekal is an Associate Professor in Mining
Engineering at the University of Wollongong, Australia. He
has both a Master's degree in Mining Engineering (India),
and a PhD in Mining, Mineral and Materials from the University of Queensland. He has accumulated over 26 years
of research experience working at several research organisations, such as the National Institute of Rock Mechanics
(India), the Centre for Mining Technology and Equipment
(CMTE), Australia, the Division of Mining, Mineral and
Materials, University of Queensland, and the CRC Mining and exploration and mining division of CSIRO. He is associated with world-renowned research organisations
such as the Norwegian Geotechnical Institute (NGI), Norway; the National Institute
of Advanced Industrial Science and Technology (AIST), Japan; the Sustainable Mining
Institute (SMI), and Julius Kruttschnitt Mineral Research Centre (JKMRC), Australia.
He has executed several national and international projects and has patents in advanced
cutting technology in USA, Germany, South Africa and Australia, and has published
several international journal and conference papers and many reports. Dr Karekal's
area of expertise includes advanced characterisation of rock, numerical modelling, fluid
flow problems, rock cutting, acoustic emission and micro-seismicity, and geophysics
and radar applications to rock mass fracture detection (e-mail: skarekal@uow.edu.au).

Chapter 1

Highwall mining: world scenarios

1.1 INTRODUCTION

There has been a great need to substantially increase the coal production in the coming decade to meet the demands of the growing economy of several countries. Due to vast development in opencast mining machinery and blasting technology, opencast mining is now planned at a greater depth. As a result, final highwall height is increased while the final highwall slope angle is reduced. Final highwalls in open-pit coal mines can form a starting point for other mining methods, such as highwall mining, which is basically a technique utilised after the open-pit portion of a reserve has been mined, sometimes prior to the introduction of underground mining (Seib, 1993). This technology enables the extraction of some thin coal seams, even less than 1.0 m, otherwise un-mineable with other existing technologies (Newman and Zipf, 2005).

Coal is the most important and abundant fossil fuel. It alone contributes to about 58% of India's energy needs. Despite an increasing trend to look for some eco-friendly resource of energy, coal continues to dominate the energy scenario and more than 80% of the total coal production comes from the opencast mines (Jha, 2010). A large amount of coal is also being imported from various countries, mainly for the energy sector. Unfortunately, many open-pit mines in India are reaching their pit limits. Existence of surface dwellings mostly limits the expansion of currently running opencast mines. Sometimes the overburden is so high that coal extraction becomes uneconomical due to enhanced stripping ratios. In order to extract the locked-up coal and to extend the life of some of the open-pit mines that have reached their pit limits, highwall mining was introduced in India recently (Loui et al., 2011). In India a huge potential for new coal mining technologies is envisioned to up the ante on much-needed coal demand for the energy sector. Highwall mining involves lesser capital but can achieve good production rate and economy. Seen as a proven technology in USA, Australia and some other parts of the world, it was not a difficult decision for Indian mining companies to try this technology. However, Indian coal geology is complex compared to some other parts of the world. The occurrence of thick and multiple seams, and some at close proximity, with frequently varying roof conditions etc., makes mining difficult and challenging.

1.2 OPERATION AND NEW DEVELOPMENTS

Highwall mining is an important surface coal mining method in many countries such as US, Australia and Indonesia. In US it accounts for approximately 4% of total US

Figure 1.1 Auger mining single and twin circular cutter heads.

coal production (Zipf and Bhatt, 2004). The technology originated in USA in 1970 and has seen widespread use since 1980. However, in Australia, highwall mining was introduced in 1990 and since then has been applied in about 40 pits at 17 coal mines. It is estimated that about 20 Mt of coal have been produced with continuous highwall mining systems (Christensen, 2004). Apart from the considerable successes, operations in Australia have experienced some problems including persistent minor roof falls to major roof failures (Christensen, 2004).

Highwall mining operation involves driving a series of parallel entries using remotely operated continuous miners with an attached conveying system into a coal seam exposed at the highwall. These parallel entries (web entries) are separated by web pillars of pre-designed width, and remain unsupported, unmanned and unventilated throughout their life. The technology is the modified version of the earlier auger mining widely practised in USA and Australia (Figure 1.1). There are two types of highwall mining systems, one being the Continuous Highwall Miner system (CHM) and the other being the auger mining system. CHM produces rectangular entries, usually of the typical widths 2.9 m and 3.5 m, while augur miners cut the coal seam in a circular manner with diameters of 1.35 to 1.8 m. The cutter-head could drive in an individual circular excavation or twin circular excavations. The first version of highwall miner was auger type which later modified to a CHM system with fully automated navigation control system. These machines can produce at an average production rate varying between 0.5 and 1 million tonnes per year, depending on the geology and the site conditions. The evolution of highwall mining started in the mid-1940s when surface blast-holes were made horizontally to extract coal from the base of the highwall (Zipf and Bhatt, 2004). By the 1950s the concept was well utilised as contour mining in hilly areas of coal fields. Maximum penetration depth that was achieved ranged from 60 to 100 m. Efforts were then made to increase the capacity and thus the augers started growing in size and power. But modified augers too were having shortcomings as detailed below:

- It is generally restricted to a small depth of penetration (<130 m) and it could be less depending on the strength of the coal; it also has to exert more power at increased depth.
- It does not have a haulage system for a longer distance and for coal seams of varying dip and roll.

- It operates at a fixed height and is thereby restricted to one thickness in varying-thickness coal seams; it is therefore not flexible and limits the coal recovery to about 30–40% in those conditions.
- It has low production rate compared to continuous miner systems such as CHM.
- Its inability to negotiate dip of the seam and diminishing power with increase in depth are other limitations.

To overcome these issues new R&D efforts ultimately led to the development of the present day CHM system. The CHM system has the following characteristics though the specifications differ from company to company:

- The system which adopts continuous miner has greater depth of penetration (about 500 m), depending on the seam dip and the conveying system adopted for conveying the debris, either traction type conveying system or push-beam type conveying system.
- Unlike the auger system, the coal fragment size is unaffected by the depth of penetration and fragment sizes depend on the lacing of picks on the drum and the power requirements.
- Since continuous miner has a cutter-head boom that can move up and down to adjust to the variable cutting heights of the coal seam, it is able to extract or recover more coal to nearly 60–80%, depending on the geological conditions. The miner is also able to negotiate changes in the dips and rolls of the coal seams.
- It can achieve higher production rate up to 0.8 to 1.0 million tonnes per year, penetrate to higher depths, and cut stronger coal.
- Depending upon the conveying traction system, the coal seams can be up-dipping (+5 deg.) and down-dipping to −15 deg.

There are three types of fragmented coal conveying systems for the CHM, namely the highwall mining system developed by Addington Resources over a period of three years, beginning in 1989 (Figures 1.2 and 1.3), the second one being the Archveyor highwall mining system, and the third being the push-beam conveying system.

ADDCAR has been taken over by UGM Group, Australia and is known as UGM ADDCAR Systems LLC. It is based in Ashland, Kentucky, USA. The company has 25 years' long experience of operating highwall mining systems in the coalfields of Australia, USA, South Africa and India and so far has produced in excess of 100 Mt of coal. It is the leader in maximising resource recovery and cost effectiveness for the clients. The ADDCAR system consists of the following components (Sartaine, 1993):

- A belt conveyor car. The launch vehicle remains stationary during the mining cycle and is used to add the cars and receive the coal from the cars. When the hole is completed, the continuous miner is backed out onto the launch vehicle and the entire system is moved over and aligned with the next hole.
- Continuous miner that fragments the coal and loads it on the cars. The head of the continuous miner is raised or lowered as the drum rotates allowing continuous miner to mine a variety of seam thicknesses.
- Belt conveyor cars that are added as the excavation progresses. Each car is 12 m long and individually powered for self-mobility.

Figure 1.2 ADDCAR highwall mining system (Courtesy: Miller Brothers Coal LLC, Kentucky, USA).

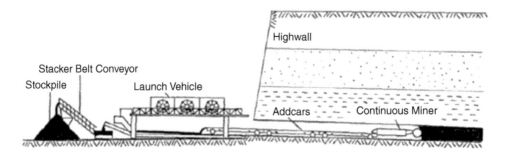

Figure 1.3 Working system of ADDCAR miner (Sasaoka et al., 2015).

- Stacker conveyor to stack the coal.
- Electric power system that receives the power from the power grid or the generator.
- Automation and navigation system to precisely control the highwall miner.

ADDCARs have traction belt cars for conveying broken coal. Hydraulically powered push arms provide thrust in addition to the tractive effort of the miner. The tractions offered by the cars facilitate entries to be driven to a longer distance into the highwall. These ADDCARs are progressively added, one in every 7.5 m, with flameproof and intrinsically safe multiple-drive motors from the launch vehicle as the face is advanced by the continuous miner. The system can handle 0.97 m to as high as 5.2 m in thickness with changing cutter-head of the continuous miner. The updated specifications can be found from their manufacturing brochures. There is a good navigation system which has the ability to negotiate dips and rolls in the seam. The built-in remote monitors provide the operator with the visual information necessary to control the system.

The Archveyor system is another type of coal conveying system from the continuous miner (Figure 1.4). The Archveyor chain conveyor is a load-out vehicle that transports coal and trams the machine system itself. The coal is conveyed through

Figure 1.4 Archveyor system (Sasaoka et al., 2015).

Figure 1.5 Schematic view of Caterpillar highwall miner and push beam conveyors (Hawley, 2008: www.prweb.com/releases/2008/06/prweb992884.htm).

Archveyor carriers to the loading point. As the continuous miner advances, the Archveyor advances to the miner once in every 1.85 m.

Caterpillar, USA has developed a push-beam auger conveying system which fits to the continuous miner. It is believed that the system can penetrate to about 350 m into coal seams of thickness ranging from 0.8 to 4.8 m, and is capable of handling seam dip up to 8°. It has flexibility in vertical direction and stiffened movement in lateral direction. The length of the push-beam conveying system can be extended on a specific order basis. Figure 1.5 shows the Caterpillar highwall miner, and the conveying system and locking mechanism for the push-beam conveying system. More details on the specifications can be obtained from the manufacturer. Currently Caterpillar enjoys the bulk of the market share in continuous highwall mining systems across the world.

During highwall mining operations safety precautions are to be taken against any unforeseen accidents such as:

- Accidents from the rock falls from the highwall or slope failure.
- Roof and rib falls at the portal.
- Crushing accidents caused by mobile equipment being operated in a confined space with poor visibility.

- Maintenance related injuries such as working on the launch pad, cranes lifting-related accidents, cutting and welding.
- Strain on the work force.
- Tripping and falling injuries.
- Risk and risk assessment due to a variety of hazards.

Highwall mining is generally adopted in the following scenarios:

- Where open-pit reaches its pit limit and further expansion is not viable – multiple and multi-pass highwall mining is possible and has been demonstrated at the end user mine site.
- Where the roof conditions are good, moderately flat seams, competent coal seam, stable highwall and strata is free from geological disturbances and presence of stone bands in the seam.
- In thin seam conditions where open-pit/opencast/contour mining is not viable, highwall mining can be adopted with trenches – again multiple-seam highwall mining is possible by digging deeper trenches.
- In old abandoned mines, highwall mining can be adopted to extract high-value coal.
- Depending on the stability of the slope, highwall mining can be adopted in situations where transition from surface mining to underground mining is probable.
- With highwall mining, selective mining can be done of high-value coal seams, or on the blocks that are free from major faults or local geological anomalies.

Some of the benefits include:

- In highwall mining, movements of equipment can be done easily as it is mobile and is used on the surface unlike in underground mining methods where movements of equipment are limited due to space constraints.
- Highwall mining offers greater safety to men and machinery, as it is operated remotely without men involved in the highwall entries and people are not exposed to hazardous conditions such as dust, fumes, gas, roof falls, irrespirable atmosphere, etc.
- Highwall mining involves low capital and low development costs, which makes the system much cheaper compared to underground mining methods such as longwall mining.
- In some mining conditions, highwall mining provides higher coal recovery for the capital expenditure; thereby it offers higher productivity and the cost per tonne is much lower compared to underground coal mining methods for winning of coal.
- It is a clean coal technology, does not generate vast overburden waste dumps like typical open-pit mines and produces coal with practically no mix-up with waste rocks.

1.3 INTERNATIONAL STATUS

The current CHM highwall mining technologies originated in USA in the mid-1970s. The advancement of the CHM and its application became quite popular in US following the auger mining system. Highwall mining approximately accounts 4% of total

Figure 1.6 Highwall collapse resulting in fatality (Zipf, 1999).

USA coal production (Zipf, 1999). In USA, the highwall mining technology was successfully implemented at Kennecott's Colowyo mine as a result of careful geotechnical and pre-production planning, and close interactions between the mine personnel and the highwall mining equipment operators (Vandergrift and Garcia, 2005). A total of 1.5 million tonnes were extracted with an average monthly production of around 100,000 tonnes. In some cases they experienced reduced penetration resulting from seam undulations and a rock parting. In eastern USA, two main factors that affected highwall stability were geologic structure and pillar stability (Zipf, 1999). Failure along the near vertical fractures in the rock can lead to large rock falls from the highwall as shown in Figure 1.6. Though backfilling of highwall mined entries is yet to be tried anywhere in the world, researchers have shown that better production and productivity can be achieved in highwall mining with backfilling, which would further increase the stability of the highwall (Wagner and Galvin, 1979; Afrouz, 1994; Matsui et al., 2000).

There have been successes and failures in highwall mining resulting from the geology and web pillar design. One of the highwall mining accidents reported by MSHA, USA (MSHA, 2000) in 2000 was a result of the unexpected fall of a large vertical plate of overburden strata, approximately 81 m in length and up to 3.65 m in depth, detached from a near vertical pre-existing fracture parallel to the highwall. It fatally injured one front-end loader operator who was working to move coal from the stockpile at the highwall miner. The fall was believed to be initiated by the large deformation of the mine pillars under it. An overall safety standard of highwall mining was found to be very good by MSHA, on par with open-pit mining.

The highwall mining method was introduced in Australia in early 1991. Since then at least 17 mines have used this method of mining spanning over 40 open-pits. Experience of highwall mining at four Australian mines, especially the geotechnical issues affecting mining performance, can be found in the paper by Shen and Duncan

Fama (1997). In Australia, ADDCAR Systems produced an average of 75,000 tonnes per month and reached a peak of 135,000 tonnes per month. These were achieved in coal seams of thickness ranging from below 1.0 m to 2.5 m (Sartaine, 1993). The short review of four highwall mining operations at Oaky Creek, German Creek, Moura and Ulan mines is given below (Shen and Duncan Fama, 1997).

- Oaky Creek mine and German Creek mine had seam thickness between 2 and 3 m. It had relatively weak coal [Uniaxial Compressive Strength (UCS) of about 3.3 MPa], weak roof (layer of weak mudstone to moderately strong siltstone and sandstone) and occasional soft floor. During mining, frequent roof falls occurred in the form of bed separation, slabbing and block falling from the cross joints. Occasional pillar spalling was observed. It was attributed that 66% of the time was spent in retracting and re-entry operation due to roof falls.
- Moura mine had seam thickness between 2.2 to 4.8 m. It had average coal strength (UCS of 4.8 MPa), moderately strong roof (thinly bedded or laminated mudstone with average UCS of 25 MPa) and had sub-vertical joints. It also had moderately strong floor. Rock falls of 0.1 to 0.3 m thick were encountered. However, roof falls were far less frequent compared to Oaky and German Creek mines. As a result, coal was left as roof layer with 0.3 to 0.5 m thickness. In some cases, the coal roof was not stable due to its low strength.

  a) In pit 17DU, roof failure extended to a height of 2–3 m, believed to have been caused by the convergence of three adjacent entries, and the continuous miner was trapped.
  b) In pit 16BL, panel pillar collapsed due to roof failure reducing W/h ratio of the pillar, and trapped the continuous miner.
  c) In pit 17DU, panel pillar collapsed after mining was carried out, with no damage.

- Ulan Mine had relatively strong coal (UCS of 35 MPa and cubic mass strength of 10–15 MPa) and strong roof (strong intact claystone of UCS 43 MPa). The floor was competent siltstone. Occasional roof falls, however, intensified towards the northern end-wall where it had free surface due to another strip cut. When the entry was at a distance of 40 m from the free end-wall, roof falls occurred at a penetration depth of 225 m, trapping the miner temporarily. In good rock mass condition the penetration depth of about 500 m was successfully achieved in that mine.

In Indonesia, about 70 million tonnes of clean coal is annually produced and 99% of the total production comes from open-pit mines. Highwall mining has been pursued by the Japanese under the joint research work conducted between Kyushu University, Japan, Institute of Technology, Bandung (ITB), and the Indonesian Institute of Sciences (LIPI), Indonesia (Matsui et al., 2001). Their aim is to extract locked-up coal in their highwalls with the help of backfilling technology. Currently in Indonesia, only the auger mining system has been used at a few mines (Furukawa et al., 2009). The reason for implementing the auger mining system in Indonesia is that their coal measure rocks are generally very weak, comprising of shale, mudstone or siltstone, showing excessive slaking behaviour leading to deterioration of their mechanical properties, and the weak soft floor tends to sink or slip the continuous miner. Further, the rectangular shaped excavation by the CHM system may cause some span instability, as opposed to circular shaped excavation by auger mining.

In Thailand, highwall mining was implemented in the EGAT Mae Moh open-pit coal mine (Sasaoka et al., 2013). The mining depth varied and the maximum depth encountered was about 260 m from the surface. This mine was geologically faulted and as a result of which, the mine had left a 200 to 300 m boundary of rock block including coal seams in front of a fault to prevent any sliding of the slope along the fault plane. The 200–300 m boundary coal was to be partially mined using highwall mining technology; otherwise it would have been locked-up in the buffered zone. This established the fact that highwall mining could be applied in a faulted region with in depth analysis.

In India a huge potential for new coal mining technologies is envisioned to up the stake on much-needed coal demand for the energy sector. Some Indian open-pit mines are reaching their pit limits. In many cases the overburden becomes so high that coal extraction becomes uneconomical due to high stripping ratios. Highwall mining offers some benefits to extract locked-up coal at the highwall. It was introduced in the year 2010–2011 and is successfully operating after some initial difficulties due to considerable dipping of coal seams. Indian coal geology is complex compared to some other parts of the world. The occurrence of thick, thin and multiple seams (some at close proximity), seams with frequently varying roof conditions, seams with varying cross-gradient and dips, seams with varying strengths, and sandstone intrusion etc., make mining difficult and challenging.

The first highwall mining operation in India was started in December 2010 at the end user site of Opencast Project (OCP-II) at Ramagundam area of Singareni Collieries Company Limited (SCCL), India. Extraction operation was carried out by Advanced Mining Technology Private Limited (AMT) using ADDCAR Highwall Mining System of USA. The original highwall mining design was provided by Agapito Associates Inc, USA. It was subsequently redesigned by CSIR-Central Institute of Mining and Fuel Research (CSIR-CIMFR), India. AMT and SCCL received approval from the Directorate General of Mine Safety, India to carry out highwall mining. Extraction has been completed at OCP-II and is now midway excavating coal at another nearby open-pit mine called Medapalli Opencast Project (MOCP). Until today, operations at these two open-pit mines have been successful after some initial difficulties. The slope monitoring was carried by AMT and CSIR-CIMFR. No subsidence or any displacement of the slope was noticed at OCP-II highwall mining site. Extraction was carried out through single and multi-pass entries in multiple-seam operations. However, not much monitoring was carried out to validate the results. At the second SCCL site, namely Medapalli Opencast Project (MOCP), two panels with multiple seams (4–5 seams) were successfully extracted including one thick seam, which was extracted in two passes. The major geotechnical issues encountered included frequent roof falls from weak clay and shale layers, steep gradient, hard coal with stone intrusions, water percolation, etc. In 2011, highwall mining started at Sharda Opencast Mine of South Eastern Coalfields Ltd. (SECL) using a trench mining concept with Caterpillar highwall miner to target thin coal seams (1.0–1.4 m thick). Here also after some initial difficulties, the highwall mining is running successfully.

There are about 25 coalfields extracting coal from various underground and opencast mines. All such coalfields have multiple extractable seams. In many instances, to the tune of about 20%, these seams occur contiguous to another seam or section (within 9 m of vertical parting). There also exist thick seams in almost all coalfields, which may require multiple-pass or sections to extract by highwall mining. Some coalfields even have over 30 workable seams. A cursory glance at the scope of highwall mining indicates that almost every coal mining area, with existing opencast extractions, provides opportunity for highwall mining.

Highwall mining is also under consideration in a few other countries in Europe and Canada.

Chapter 2

Geological characterisation for highwall mining operations

2.1 INTRODUCTION

In this chapter pertinent methods on geological characterisation of the rock layers in any highwall mining site are elaborated with the help of a case example from the Godavari coal basin, situated in central India.

Highwall mining projects are often situated in sedimentary coal basins. Therefore the morphology of the surface terrain, type and structures of coal and coal-bearing strata, physical, mechanical and hydro-geological characteristics of the strata, must be considered in detail in the natural state and also under mining stress conditions to understand the possible behavioural changes. In general, geological structures, size/scale, stress/strain and strength of rock/rock mass play an important role in the highwall mining design. Structural properties of sedimentary rock formation can never be overestimated owing to the fact that failure is often governed by the weakest characteristics in the rock mass. It is often necessary to consider geological characterisation over a regional scale and then focus more on the detailed variation in the local geology around the highwall mining blocks. A proactive approach is needed to know the details of the roof and floor layers of the coal seam targeted for highwall excavations, emphasising structural features such as bedding planes, layer thickness, layer shear strength, cleats and their orientation, igneous intrusions, dykes, faults, folds, joints and other forms of discontinuities. Furthermore, geo-mechanical (physico-mechanical) and hydrological properties of coal and the coal-bearing strata play an important role in structural characterisation. The main purpose of geological characterisation is to provide adequate geotechnical knowledge of a mining area so as to identify the regions of similar geotechnical properties and to estimate *in situ* response of excavations under geological conditions that exist at a given site of interest.

In view of this, a hydro-geo-mechanical framework is necessary to provide an integrated skeletal structure for site characterisation of highwall mining. In common phraseology *"garbage in, garbage out"*, site characterisation resolves data uncertainty and provides more useful and meaningful input to the tools used in the assessment of span stability, pillar stability and panel stability in the case of single, multiple and multi-pass highwall mining excavations and engineering judgement on the design process.

In this chapter, a general geo-mechanical framework (Figure 2.1) has been discussed by considering geology, geophysics, hydro-geology, geotechnical properties,

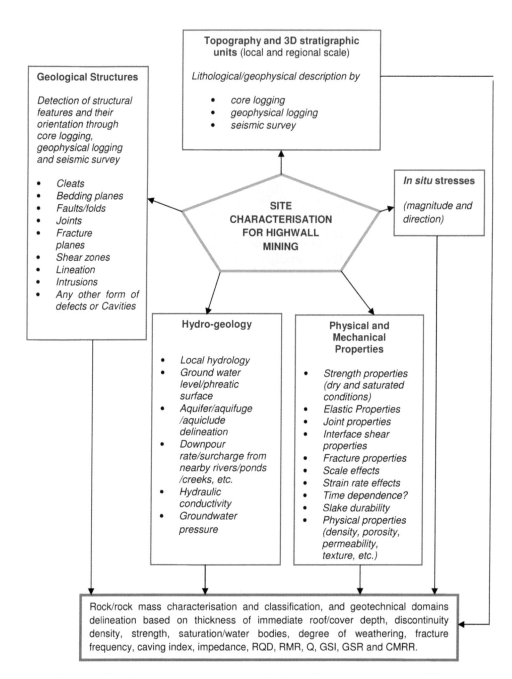

Figure 2.1 Site characterisation components for highwall mining excavations and slope.

geological structures and *in situ* stresses that affect the stability of highwall slope and highwall mining excavations. The importance of these aspects in the design of highwall mining excavations is highlighted using case studies of site investigations conducted at Godavari valley in central India.

2.2 FRAMEWORK FOR SITE CHARACTERISATION

2.2.1 Topography and 3D stratigraphic units

Surface topographic information is imperative in highwall mining excavation and high-wall slope stability assessment as the magnitude and direction of *in situ* stresses are controlled to a certain extent by the surface topography. Depth of cover over the coal seam of interest where highwall mining excavations are to be driven depends on the surface topography as vertical stress increases with the increase in the depth of cover. Any abrupt variation in the topography has to be taken into account, in addition to topographical changes due to long-term processes such as erosion, drifts, etc. A topographic survey could also assist in fault interpretation.

A detailed 3D stratigraphic model of the site of interest is a prerequisite, which can be developed by considering core logging and geophysical logging data. As a rule of thumb about 16–20 boreholes per square kilometre provide a good estimate of the geology. However, spacing of the borehole depends on what scale of information one would like to seek. At the end, it all depends on the benefit one gets against the cost incurred in drilling. Signature pattern analysis of wireline geophysical log data has been proved to be very useful in describing the petrographic variations in the rocks, like variation in grain size or composition within each bed. From the geotechnical point of view, pattern recognition of log data helps in delineating any weak planes which could influence the rock caving process. The available advanced statistical techniques/packages, such as Self Organising Map (SOM) or LogTrans, can be used in classifying the coal and coal-bearing rock units based on geophysical parameters. Figure 2.2 shows one such example where lithology of the end-user site has been interpreted using geophysical logs of Ramagundam area, India (Shanmukha et al., 2015).

A borehole acoustic scanner can also be used in delineating fractures or bedding planes. These fracture planes can then be imported into Discrete Fracture Network software to simulate rock mass structure which then can be used to predict the *in situ* response of rock mass using numerical tools. Figure 2.3 indicates regional consistency and near parallelism of the various beds making up the Permian sequences. These beds were classified based on geophysical sonic logs. Sonic velocity maps can be useful in providing first-hand information of the various beds and their relative strength values for undertaking 3D numerical modelling of highwall mining and highwall slope stability analysis.

Such characterisation enables classification of the geological sequences based on the uniaxial compressive strength values.

2.2.2 Geophysical studies at Godavari valley in India

Godavari valley is situated in the central part of India, where highwall mining was proposed at two open-pit mines operated by Singareni Collieries Company Ltd. (SCCL). Empirical studies were undertaken to establish an exponential relationship between primary wave velocity (V_p), obtained from geophysical sonic logs, and UCS determined on core samples of sandstones from all parts of Godavari valley of the proposed highwall mining basin in India (Shanmukh et al., 2015). The middle Permian Barren Measure (BM), having a thickness of 200 to 250 m, is resolved into upper and lower sequences dominated by UCS of relatively lower and higher values of upper

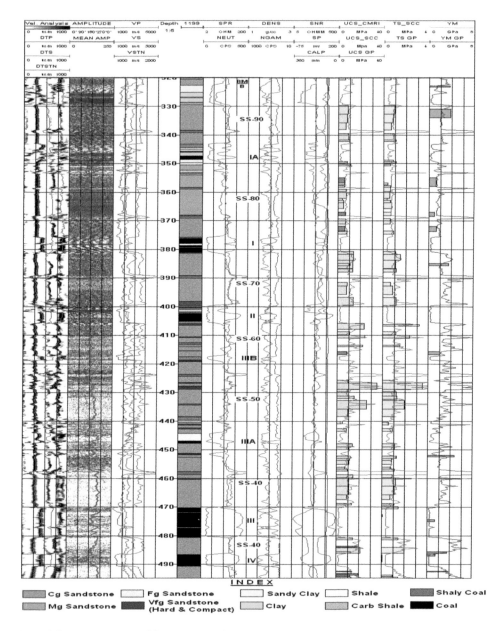

Figure 2.2 Interpretation of lithologies and geotechnical properties of Barakar Formation using geophysical logs, borehole RG-1199, Adriyala longwall block, Ramagundam area.

and lower sequences, respectively. Early Permian coal-bearing Barakar Formation of 200 m thickness is also resolved into upper and lower sequences dominated by UCS of relatively higher values compared to BMs. Sandstones of BM are poorly sorted, less compact, less cemented and more friable than the sandstones of the Barakar.

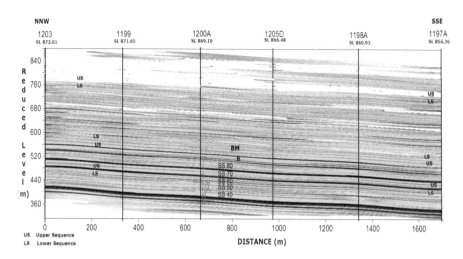

Figure 2.3 V_p derived UCS map of Permian sequences of Adriyala longwall block, Ramagundam, India. [BM indicates Barren Measures and B indicates Barakar coal-bearing Formation; US – Upper Sequence; LS – Lower Sequence; BM – Middle Permian Barren Measure Formation; B – early Permian Barakar Formation (coal-bearing); IA, I, II, IIIB, IIIA, III, IV coal seams; SS-90, SS-80, SS-70 Nomenclature of overburden strata].

Sandstones of the BM are also more kaolinised and are prone to weathering into clay bearing rocks, which seem to influence UCS values. Competency of Permian sequences increases in the descending order.

2.2.3   Geological structures

Structural discontinuities and their orientation govern the strength and deformation of rock mass. Structural discontinuities include cleats, bedding planes, faults/folds, joints, fracture planes, shear zones, lineation and lineaments, intrusions and any other form of defects or cavities, and these are considered as stress raisers in the rock mass. These features act as a weakest link in the rock mass affecting the stability of structures. They also increase the hydraulic conductivity and permeability of rock mass. These geological discontinuities often dictate mine design layouts. Seismic survey (2D or 3D) can be undertaken to delineate the geological features such as faults, intrusion, folds or major discontinuities, etc. However, time, cost and the value (benefit) need to be assessed before undertaking any such investigations.

2.2.4   *In situ* stress

The *in situ* stress field is generally considered to be important as these values are useful in the design layouts of highwall mining. In general, highwall mining layouts, face conditions and caving behaviour issues like roof sag, guttering, rib spalling which necessitate excessive support requirements can be related to the *in situ* stress state.

The buckling of roof under high horizontal stress, combined with bending due to vertical load, aggravates roof failure and controls the span stability. Generally, highwall mining excavations driven perpendicular to the major principal stress directions tend to delaminate the roof and aggravate the roof failure whereas the highwall mining excavations driven in the direction of major principal stress direction are less prone to roof damages. Local faults could also change the local *in situ* stress direction. In such situations, the focus can then be laid on geological structures in the design of highwall mining for optimum reserve exploitation.

2.2.5 Physical and mechanical properties

For undertaking any highwall mining design, one must consider physical and mechanical properties, including the property change under dry and water saturated conditions, especially strength change, slaking durability of immediate roof and floor of highwall mining excavations, elastic properties (Young's modulus and Poisson's ratio), joint properties, fracture properties, scale effects, strain rate effects, time dependent properties, density, porosity, permeability, texture, etc. that characterise the constitutive behaviour of rock/rock mass. These properties must be determined precisely without any bias, with proper sample preparation and use of calibrated testing machines. For testing methods, sample preparation and data recording one can refer to appropriate ASTM (American Society for Testing and Materials) or ISRM (International Society for Rock Mechanics) testing standards.

2.2.5.1 Physico-mechanical property determination in Godavari Basin

Some of the distributions of elastic and mechanical properties for various overburden layers of some of the areas of SCCL (Adriyala and Kakatiya Khani sites of the Singareni Collieries Company Ltd. of India) are shown in Figures 2.4 to 2.8 (Karekal et al., 2008). Figures 2.9 and 2.10 are the reduction of strength due to water saturation in sandstones. One can infer from such results that water saturation due to the presence of groundwater can have greater influence on the reduction of strength of sandstones.

2.2.6 *In situ* strength of coal

Coal strength and deformability are significantly affected by the composition of the coal and the presence of flaws/defects like cleats or fracture planes. The scale effect is a dominant feature of rocks and coal behaviour – as with the increase in size/volume of the rock/coal, its strength would decrease due to the probability of encountering longer cleats and cracks. Figure 2.11 and Figure 2.12 show the appearance of coal and rock at different scales respectively. As indicated in the figures, scale effect governs the strength to certain volume and is believed to be independent of volume beyond certain representative volume of coal and rock mass.

Various scale effect equations that are available in the literature were assessed. Figure 2.13 shows the percent reduction in strength with increase in size for coal, coal measure rocks and hard rocks (Hoek and Brown, 1980; McNally, 1996). It is evident that Medhurst's (1996) formulation significantly reduces the strength of

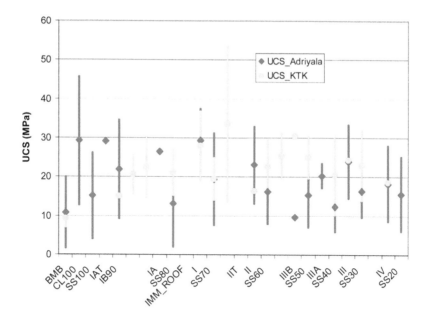

Figure 2.4 Uniaxial strength distributions (with mean and standard deviation) across the stratigraphic units of Adriyala and KTK sites.

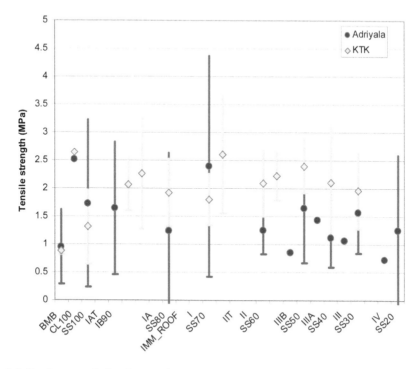

Figure 2.5 Tensile strength distributions (with mean and standard deviation) across the stratigraphic units of Adriyala and KTK sites.

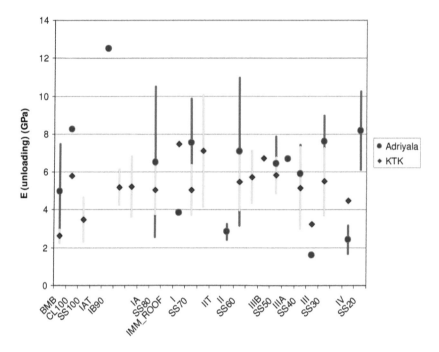

Figure 2.6 Young's modulus distributions (with mean and standard deviation) across the stratigraphic units of Adriyala and KTK sites.

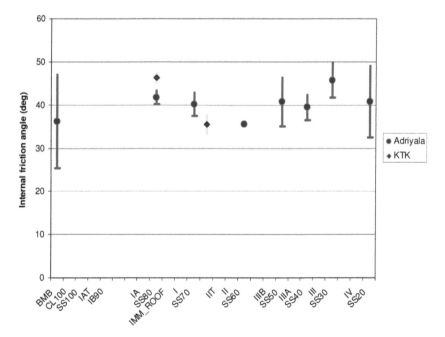

Figure 2.7 Internal friction angle distributions (with mean and standard deviation) across the stratigraphic units of Adriyala and KTK sites.

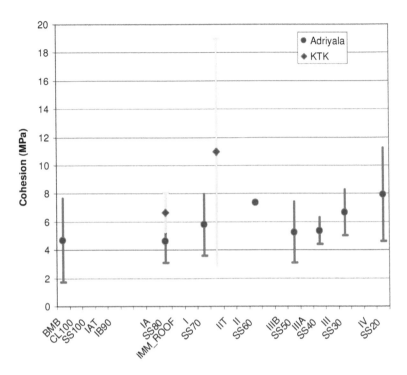

Figure 2.8 Cohesion distributions (with mean and standard deviation) across the stratigraphic units of Adriyala and KTK sites.

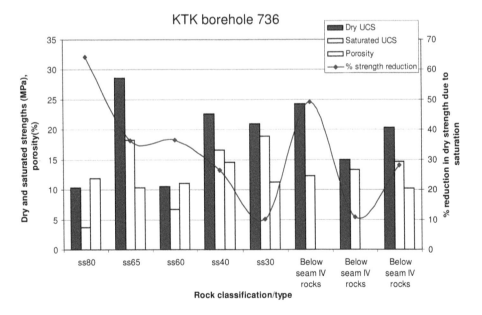

Figure 2.9 Dry and saturated strength (UCS) distributions and % reduction in dry strength due to water saturation along with the porosity values across the stratigraphic units of KTK (borehole 736).

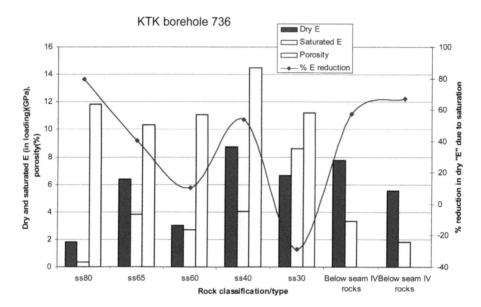

Figure 2.10 Dry and saturated Young's moduli distributions and % reduction in dry Young's modulus due to water saturation along with the porosity values across the stratigraphic units of KTK (borehole 736).

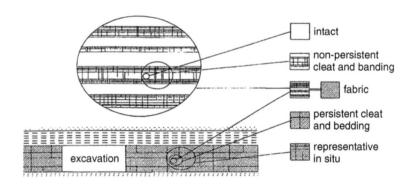

Figure 2.11 Idealised diagram showing the appearance of coal at different scales (Trueman and Medhurst, 1994).

coal with increase in size compared to McNally's formulation (1996). The hard rocks exhibit relatively less reduction in strength when compared to coal measure rocks. These formulations can be used to estimate the strength values for large size samples (equal to the mesh size in numerical modelling) and can be used as a starting point for calibrating numerical models.

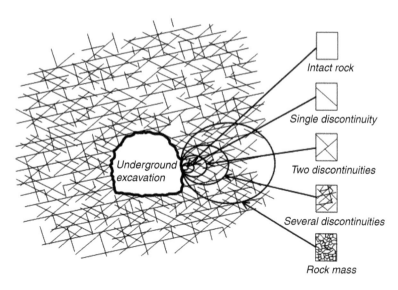

Figure 2.12 Diagram showing the defect inclusions at different scales in hard rocks (Hoek and Brown, 1980).

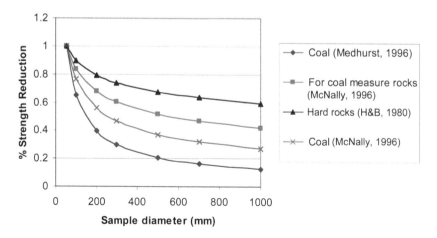

Figure 2.13 Predicted UCS strength reduction with increase in size using scaling laws developed by various authors.

2.2.6.1 *In situ strength test at Godavari Basin*

In situ strength tests carried out in seam I and seam III at GDK-10A Incline by National Institute of Rock Mechanics (NIRM), India (2008) provide some indicative strength values of coal. A series of *in situ* strength tests carried on 30 cm × 30 cm × 30 cm blocks showed a minimum measured value of 3.5 MPa, and the maximum was 10.5 MPa. The stiffness values for seam I were measured between 30,000 and 40,000 kg/cm². The minimum *in situ* strength of coal for seam III was 4.74 MPa (when corner failed)

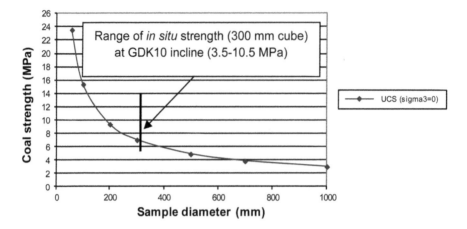

Figure 2.14 Scale effect prediction based on Medhurst's equation (1996) with Lab. UCS~23.7 MPa.

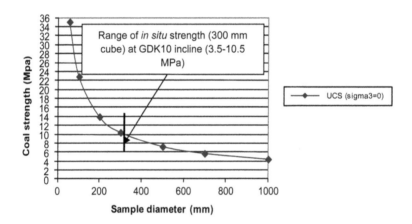

Figure 2.15 Scale effect of prediction based on Medhurst's equation (1996) with lab. UCS~35 MPa.

and the maximum measured was 19.27 MPa. The measured average strength was 13.2 MPa. Coal from seam III was found to be stiffer than seam I, with stiffness value ranging between 50,000 and 70,000 kg/cm^2.

Scale effect studies on coal were undertaken based on Medhurst's formulation (Medhurst, 1996). Laboratory tests carried out by CSIR-CIMFR on NX size cored coal samples of seam I revealed the compressive strength in the range of 215 to 234 kg/cm^2. These correspond to the mid brighter value of coal of ~275 kg/cm^2 tested by Medhurst (1996) and therefore the scaling values (for "m" and "s") developed by Medhurst (1996) were used for calculations. Figures 2.14 and 2.15 show variations of UCS as a function of size/diameter for two cases (case-I: 23.4 MPa and case-II: 34 MPa for NX size samples). The strength (UCS) reduces significantly with the increase in size.

The vertical bar in the figures shows the range of observed *in situ* strength values (on 30 cm × 30 cm × 30 cm block) at the same site.

2.2.7 Hydro-geology

Special attention needs to be given to hydro-geological properties as they adversely affect the stability of geotechnical structures and have large influence on the stability. The presence of water creates pore pressure, thus reducing the effective normal stress. It also reduces the friction and cohesion values but often increases the joint aperture by washing away the infillings in the joints.

In common with other rock masses, water flow in coal measure rocks takes place dominantly through defects i.e. joints, bedding planes, faults and cleats. Water washes away the infillings in the joints and increases the joint aperture. Typically, in the Permian coal measures, the coal seams are the major aquifers. Hydraulic conductivities for coal are in the range of 0.01 to 1 m/day, with specific yields of about 2%, and elastic storage of 0.0002 (Seedsman, 1996). The actual values depend on the nature of cleating and also on the *in situ* stress conditions. Fine-grained units such as mudstones, shales and clays act as aquicludes.

Inflow of water could be associated with the severe face falls (Xiao et al., 1991; Seedsman, 1996). Surface cracking due to subsidence may impact on the local hydrogeological regime by increasing drainage rates to the groundwater table. Geological structures such as joints, dykes and faults provide pre-existing pathways for water inflows and can have high conductivity. The hydraulic conductivity be enhanced due to mining activities (Seedsman, 1996). One needs a water management scheme (or constant water pumping), especially during mining down dip in highwall mining. Water in the highwall excavation affects the roadway manoeuvrability of the continuous miner and the conveying system. On the geotechnical side, water affects the weatherability of rocks and thereby reduces rock strength. Saturated sandstones have shown to exhibit reduced strength compared to their dry strength.

2.3 DOMAIN BLOCK DELINEATION FOR HIGHWALL MINING

After assessing the major elements of a geo-mechanical framework, as shown in Figure 2.1, strata characterisation and classification can be undertaken from the core log/geophysical log/seismic and laboratory geotechnical test data. More importantly for mine planning layout of highwall mining, geotechnical domains of a particular site can be set up and hazard maps can then be prepared. The geotechnical domains can be established based on immediate roof thickness distribution or the layering within the immediate roof horizon. Its thickness and strength information can be obtained from geophysical logging and estimation of geophysical strata rating. The immediate roof information is vital as thickness and properties of roof layering control the roof collapse and damage to the conveying system of the highwall miner. If the roof collapse becomes severe then there would be difficulties in pulling or withdrawing the continuous miner used for the highwall mining system, and an increased chance of burying an expensive piece of highwall mining equipment. Another factor which governs the domain delineation in highwall mining is based on geological structures such as fracture frequency, discontinuity, fault density, or strength distribution of rock mass. The third governing factor for domain delineation could be based on water retaining layers/water bodies or degree of weathering. The domain delineation could also be based on Rock Mass Rating (RMR) which is widely used in India for their roof support design. Rock mass quality index such as Q, Geological Strength Index, Coal Mine Roof Rating (CMRR) or Geophysical Strata Rating, are also widely used.

Highwall mining can become uneconomic if the roof is so weak that it collapses before the miner has been withdrawn from the hole. The RMR and CMRR have been used to evaluate potential highwall mining reserves and to identify likely unsuitable areas (Hoelle, 2003).

Once this process is completed, assessment can then be undertaken to predict site-specific behaviour of rock mass to assist with the design using engineering judgement on the prevailing conditions. This geotechnical domain characterisation and delineation, with more site-specific quantification of deformation and failure by numerical modelling, would offer great information for developing hazard maps, thereby becoming handy in assessing geotechnical risks in advance and thus facilitating in the development of management plans to control or mitigate risks in anticipated adverse situations in highwall mining.

2.4   DETERMINATION OF DATA UNCERTAINTY

Confidence in input data to a geotechnical model can be quantified using a statistical approach where a parameter is considered to lie within two-sided confidence limits (defined as upper and lower limits) for a given confidence level. Confidence limits about the population mean and the population standard deviation can be estimated on the basis of the sample standard deviation and the sample size (e.g. number of observations, tests etc.). It is obvious that, if the variability among tests (observations) is small as reflected by the sample standard deviation, the mean value can be estimated within narrow limits. However, if the variability is large (as indicated by the sample standard deviation) the mean can be estimated only within broad limits. It is obvious that a better estimate of the sample mean can be obtained with a larger number of tests (observations) than with only a few.

Uncertainty and variability in the physico-mechanical properties are inevitable. Therefore, selection of representative values from widely scattered data becomes challenging. These uncertainties in the input data contribute to model uncertainties. One of the ways to provide bounds/limits to the data uncertainty is by the use of confidence interval. The confidence interval estimate was devised to include three basic attributes of a sample (i.e. a set of laboratory tests or field measurements) (a) the most expected value (mean), (b) the variability as measured by the standard deviation and (c) the size of the sample. In this approach the estimated parameter (e.g. expected value) is bounded by two values, so-called confidence limits.

The confidence limit of a parameter can be estimated using a statistical method first published by W. S. Gosset (1908) using his method known as "Student-t Test". In this method the parameter labelled as t depends on a variable v termed as "degree of freedom" ($v = N - 1$, where N is the number of observations/tests). If $\bar{x}$ and s are the

Table 2.1 Student-t distribution (Marsal, 1987).

Confidence limit/ Degrees of freedom (N − 1)	50%	80%	90%	95%	99%
1	1.000	3.078	6.314	12.706	63.657
10	0.700	1.372	1.812	2.228	3.169
20	0.687	1.325	1.725	2.086	2.845
30	0.683	1.310	1.697	2.042	2.750
40	0.681	1.303	1.684	2.021	2.704
50	0.679	1.299	1.676	2.009	2.678
60	0.679	1.296	1.671	2.000	2.660
70	0.678	1.294	1.667	1.994	2.648
80	0.678	1.292	1.664	1.990	2.639
90	0.677	1.291	1.662	1.987	2.632
100	0.677	1.290	1.660	1.984	2.626
∞	0.674	1.282	1.645	1.960	2.576

sample mean and standard deviation respectively then the population (or real) mean is defined to lie in the range:

$$\mu = \bar{x} \pm t.s/\sqrt{N-1} \tag{2.1}$$

where t corresponds to different values of confidence intervals and v, as shown in Table 2.1.

The t distribution should be used in making confidence interval estimates when the true standard deviation of the population is unknown and estimated using s, the standard deviation estimated from the sample. The magnitude of the error involved in using the normal approximation rather than t varies with the size of the sample, becoming smaller as the size of the sample is increased ($N > 30$). However, N being less than 30 is common in routine mine geotechnical studies; for example, the number of laboratory tests on rock strength is often limited and less than 30.

For example, if the number of tests are $N = 27$ and their mean UCS $= 10$ MPa with standard deviation of 5 MPa, the Student-t Test yields a value of 2.052 for 95% confidence level that the expected mean (μ) value of UCS lies within the range:

$$10 - 2.056 * \frac{5}{\sqrt{N-1}} < \mu < 10 + 2.056 * \frac{5}{\sqrt{N-1}}$$

i.e. $\mu = 10 \pm 2.016$

or $8 < \mu < 12$

(2.2)

One needs to increase the number of tests (N) to narrow down the upper and lower bounds for the mean. Increased number of tests on the other hand may also help in lowering the standard deviation. The site investigation conducted at Adriyala Longwall Project is shown in Table 2.2, which gives estimated confidence limits (at 80%, 90% and 95%) to the various properties of Adriyala rock units, wherever possible. For some properties, the large range in the confidence level is due to a fewer number of tests.

Table 2.2 Estimated expected mean value intervals using Student-t Test distribution for various properties of some of coal and coal-bearing rock at Adriyala Longwall Project.

ADRIYALA Rock strata Mean UCS Range (MPa)	BMB		SS100		IB90		SS80	
	From	To	From	To	From	To	From	To
80% CI	10.0	11.6	11.67	18.60	12.3	31.6	11.3	15.1
90% CI	9.8	11.9	10.62	19.64	8.5	35.3	10.8	15.7
95% CI	9.5	12.1	9.66	20.60	4.5	39.4	10.3	16.1
E (GPa) (unloading)								
80% CI	4.4	5.5					5.6	7.4
90% CI	4.2	5.7					5.3	7.7
95% CI	4.0	5.9					5.1	7.9
Cohesion (MPa)								
80% CI							2.6	6.6
90% CI							1.5	7.7
95% CI							0.1*	9.2*
Friction angle (deg)								
80% CI							39.6	43.7
90% CI							38.4	44.9
95% CI							36.9	46.4
Density (kg/m³)								
80% CI	2041.4	2073.6	2134.5	2277.8	2040.1	2292.9	2115.0	2169.7
90% CI	2036.9	2078.2	2112.9	2299.4	1993.9	2339.1	2107.1	2177.6
95% CI	2032.9	2082.2	2093.3	2318.9	1946.3	2386.7	2100.2	2184.5
Poisson's ratio								
80% CI	0.2	0.2					0.2	0.2
90% CI	0.2	0.2					0.2	0.2
95% CI	0.1	0.2					0.2	0.2
Tensile strength (MPa)								
80% CI	0.9	1.0	1.26	2.198			0.9	1.6
90% CI	0.9	1.0	1.12	2.34			0.8	1.7
95% CI	0.8	1.1	0.98	2.47			0.7	1.8

*Three triaxial test data

2.5   RECOMMENDATION FOR IDENTIFICATION OF HIGHWALL MINING BLOCKS

The site investigations for identification of suitability of highwall mining blocks must consider the geological characterisation as described in the previous sections. This must be followed by the review of mine plan of the site and long-term planning. In some cases, if mine management wants to extend the open pit in any other directions other than the highwall mining side, then consideration must be given to long-term stability of the highwall mining excavations, and the logistics aspects. In the case of multiple seam highwall mining, backfilling of the pit to make a platform for driving excavations into the upper seams is necessary. The cost associated with planning and logistics must be estimated properly considering the constraints like power consumption and time factor, especially when multiple seam highwall mining is considered for operation. Sufficient contingencies must be put in place, such as overhead for delays in implementation in logistics and planning. Provision of enough budget shall be put aside for basic site investigation, second-pass investigation for detail characterisation and design, as well as for additional investigations. Detailed economic viability should be carried out, incorporating costs associated with site investigations, monitoring, site preparation, pit backfilling for multiple seam highwall mining, power, dewatering, site logistics, equipment, etc.

Among many geological features, the following are quite important for the highwall mining design and operations:

- Seam conditions – dip, rolls, thickness variations, strength, methane/CO_2 emissions, cleat partings, sandstone or hard band lenses within the coal seams, structure, groundwater, etc.
- Roof and floor conditions – strengths, presence of clay bands and shales, bedding planes and their thicknesses, roof/seam interface, structures, lithology and stratigraphy, etc.
- Highwall – slope angle, structures and formation of planar, wedge or toppling or combination of these, blast impacts and damages due to weathering, groundwater run offs, etc.
- Overburden and surface conditions – overburden thicknesses and its densities, topography, dumps, rivers, canals, villages, agriculture lands and forestlands, railway lines, monuments, etc.
- Geological structures – faults, folds, dyke intrusions, etc.

Chapter 3

Design formulations for web pillar and entry span

3.1 INTRODUCTION

For pillar design (web or barrier pillars), one needs to judge the pillar based on laboratory and field experiments, and numerical modelling, as well as empirical and experiential approaches. In-depth analysis of case studies about their failure has revealed that accurate web pillar design is the key to successful highwall mining operations. Web pillar design involves three major steps viz., estimation of web pillar strength, estimation of load on web pillar and determination of Factors of Safety (FoS).

A study of underground stone mines found that the strength of slender pillars that is W/h ratio less than two is highly variable and follows a different failure process compared to traditional square pillars (Esterhuizen, 2006). Thus, countries practising highwall mining have developed their own pillar strength formulae suitable for highwall mining scenarios.

There are several parameters that affect the span stability in mining excavations. Based on past experience, some good empirical classification systems, such as Rock Mass Rating (RMR), Q-system (Q) and Coal Mine Roof Rating (CMRR) have been devised that could be used for designing the span. There are standard charts that use classification systems to design unsupported spans, such as highwall mining excavations based on a large number of case studies, and which have been used in practice for mining and civil engineering excavations. However, rigorous numerical modelling, in conjunction with empirical adaptation, would provide further confidence in the design. Site experiences acquired during highwall mining operations would be of great benefit in validating the designs for combating roof problems.

3.2 PILLAR STRENGTH FORMULAE

There are many pillar strength formulae proposed by various authors. Among them a few have been modified and used in the design of web or barrier pillars in various case studies discussed in Chapters 4 and 6 and which are summarised below (Verma et al., 2014; Jawed et al., 2013).

3.2.1 Obert and Duval formula

Obert and Duval proposed pillar strength formula in general for square and rectangular pillars used in underground mines. The formula proposed by Obert and Duval (1967) is given below:

$$S = S_m * \left(0.778 + 0.222 * \frac{w}{h}\right) \tag{3.1}$$

where S is pillar strength (MPa), S_m is the cubic coal strength, h is the pillar height (m) and w is the width (m) of a square or rectangular pillar. In the case of rectangular pillars, the smaller dimension is taken as w.

3.2.2 Salamon and Munro formula

Salamon and Munro (1967) carried out an intensive investigation into the strength of rectangular shaped coal pillars by statistically analysing 96 intact and 27 collapsed pillar geometries in South African coal mines.

Australian mining conditions are well represented by Salamon-Munro equation. The generalised equation is given below:

$$S = K_0.h^\alpha.w^\beta \tag{3.2}$$

where,
S – pillar strength, MPa
K_0 – mass cube strength $= 7.176\,\mathrm{kPa}$
h – pillar height (m), w is the width of a square or rectangular pillar (m)
$\alpha = -0.66$ and $\beta = 0.46$.

The main reason for the popularity of the Salamon and Munro (1967) equation is that the data used was from the mines and the strength was taken as the mean strength of coal pillars, as opposed to the strength of coal specimens. The strength equation is applicable to slender pillars. For evaluation of the long-term stability of Indian coal pillars, Sheorey et al. (1986) generalised the Salamon and Munro (1967) equation as:

$$S = 0.79K.w^{0.46}/h^{0.66} \tag{3.3}$$

where K (cube strength) needs to be evaluated by the testing of specimens 30 cm in size. If no data exists regarding compressive strength of large specimens, the traditional approach is to apply a reduction factor to the USC estimated from small rock samples (Vandergrift and Garcia, 2005), in an order that ranges from 30–70% (Iannacchione, 1999).

Over the period of last three decades, there have been many pillar strength formulae proposed by various authors. Among them a few formulae that have been used in the design of web or barrier pillars are summarised below.

3.2.3 Mark-Bieniawski pillar strength formula

Mark et al. (1988) and Mark and Iannacchione (1992) proposed a new pillar strength equation which is widely known as the 'Mark-Bieniawski' pillar strength equation.

In the US, the Mark-Bieniawski pillar strength formula for slender pillars, along with tributary approach, is used to compute FoS of the web pillar and barrier pillar. It is given by:

$$S = S_m * \left(0.64 + 0.54 * \frac{W}{h}\right)$$

(3.4)

where S is web pillar or barrier pillar strength, S_m is the *in situ* coal strength, h is the mining entry height and W is the web or barrier pillar width.

3.2.4 CSIR-CIMFR pillar strength formula

The CSIR-CIMFR pillar strength equation has been developed over a couple of decades after analysing a large number of pillar stability observations from a gamut of Indian mining scenarios. The CSIR-CIMFR pillar strength equation developed by Sheorey (1992) is given below:

$$S = 0.27 * \sigma_c * h^{-0.36} + \left(\frac{H}{250} + 1\right)\left(\frac{W_e}{h} - 1\right)$$

(3.5)

where,
S is web pillar or barrier pillar strength (MPa)
H is the cover depth (m)
W_e is the equivalent pillar width (m) $= 2W$ for long pillar
W is the width of the web pillar (m)
h is the working mining entry height (m)
σ_c is the strength of 25 mm cube coal sample (MPa)

Any deviation in pillar width will lead to a reduction of pillar strength and thereby its safety factor (CSIR-CIMFR, 2008). Thus, to account for deviation in pillar width, the following correction is applied and accordingly pillar strength is computed.

$$W_d = W \pm 2dL$$

(3.6)

where,
W_d = deviated width (m)
W = web pillar width (m)
d = percentage deviation with respect to length of web cut
L = maximum length of extraction cut (m)

3.2.5 CSIRO, Australia pillar strength formula

The strength formula given below is obtained from the best linear fit to the data from Duncan Fama et al. (1999) and Adhikary and Duncan Fama (2001). It is applicable for a distance of 50 m from the portal and at least 25 m from the end of the entry. A detailed discussion on CSIRO, Australia pillar strength formula is provided in Section 6.7.2 of Chapter 6.

$$S_{\text{CSIROPART}} = 6.36 * (0.41 + 0.59\,W/h) \quad [0.5 \le W/h \le 2.0] \tag{3.7}$$

where S is the pillar strength.

Wagner (1974) proposed the equation:

$$S_{\text{WAGNER}} = 6.00 * (0.64 + 0.36\,W_e/h) \tag{3.8}$$

with $W_e = 2W$ for very long rectangular pillars of cross-sectional width W and height h.

It is noted that the slope of Equation (3.7) is much greater than that of Equation (3.8). By defining $W_e = [(0.69 + 0.44\,W/h)W]$ for $0.5 \le W/h \le 3.0$ and $W_e = 2W$ for $W/h \ge 3.0$ in Equation (3.8) the proposed CSIRO empirical formula becomes:

$$S_{\text{CSIRO}} = 6.00 * [0.64 + 0.36(0.69 + 0.44\,W/h)W/h] \text{ for } 0.5 \le W/h \le 3.0 \tag{3.9}$$

and Equation (3.8) for $W/h \ge 3.0$.

3.2.6 Hustrulid and Swanson formula

In one particular case, the Collinsville Multilevel coal mining project ("Collinsville Coal" Glencore Xstrata), the empirical relation after Hustrulid and Swanson (1977) and later modified by Wagner (1974) was utilised for pillar design. The modified formula is as follows:

$$S = S_c \left[\alpha + (1 - \alpha)\, \frac{W_e}{h} \right] \quad \text{where } W_e = \frac{2W}{(1 + W/L)} \tag{3.10}$$

where S is pillar strength; S_c is seam unit strength; W is pillar width; h is pillar height; L is pillar length and α is a constant, mainly related to opening shape. When L is large relative to width, as for highwall mining pillars, the above equation modifies to:

$$S = S_c \left[\alpha + 2(1 - \alpha)\, \frac{W}{h} \right] \tag{3.11}$$

Based on experience in Appalachia, Hustrulid and Swanson (1977) suggested a value of $\alpha = 0.78$ for a wide range of rectangular openings shapes. The static FoS for selected pillar sizes can be computed as the ratio of pillar strength to pillar load.

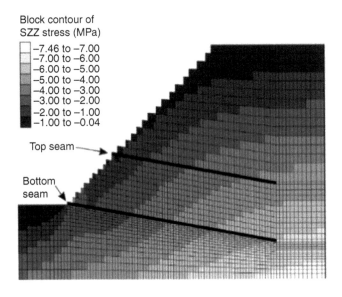

Block contour of
SZZ stress (MPa)

☐ −7.46 to −7.00
−7.00 to −6.00
−6.00 to −5.00
−5.00 to −4.00
−4.00 to −3.00
−3.00 to −2.00
−2.00 to −1.00
−1.00 to −0.04

Top seam

Bottom seam

Figure 3.1 Modelled zone (Porathur et al., 2013).

3.2.7 Load on pillars

Load on pillars is generally estimated using the tributary area theory, from the following equation:

$$P = \frac{\gamma H(W + W_c)}{W} \tag{3.12}$$

where P is the load on web pillar (MPa); γ is the unit rock pressure (MPa/m); H is the depth of cover; W is the width of web pillar and W_c is the width of web cut. However the load 'P' is usually not uniform throughout the length of web pillars due to various other factors, such as depth variation, end effect, or induced abutment stresses caused by pit or trench excavation, thus it is prudent to obtain the maximum vertical stress acting on a web pillar using numerical modelling.

In the Sharda OCP mine the seam is flat, but in OCP-II, MOCP and West Bokaro seams dip from −3° (up-dip) to 14° (down dip). Due to this inclination overburden depth varies from highwall face to final depth of penetration. Hence in order to estimate the load on the web pillar with fair accuracy a detailed numerical modelling analysis was done (Porathur et al., 2013). A three-dimensional model was prepared in *FLAC3D* software for OCP-II and MOCP site data using measured *in situ* stress data from nearby mines and the load pattern under different situations was analysed. Two coal seams were modelled as shown in Figure 3.1.

Comparison of theoretical and numerical vertical stress on the seam horizon before extraction, and the web pillar after extraction from bottom seam is shown

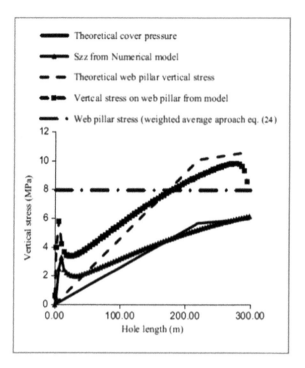

Figure 3.2 Comparison of theoretical and numerical vertical stress acting on seam horizon before extraction and on web pillar after extraction for bottom seam (Porathur et al., 2013).

in Figure 3.2. For bottom seam before any extraction, theoretical vertical stress computed from tributary area approach shows a continuously rising trend, but vertical stress computed from the model shows a steep rise near the toe of the slope to a distance of 10–15 m, and then falls before rising again with an increasing depth of penetration as well as depth of cover.

After driving of web cut to full penetration, theoretical vertical stress shows a higher stress value but the trend is the same as that of theoretical vertical stress before extraction. Nevertheless vertical stress on the web pillar shows a steep rise near the toe and a small distance inside the highwall, and then a fall in stress is observed, which rises again with increasing depth and drops down to a lower stress value on reaching the solid boundary at the end of web cut (Figure 3.2). A weighted approach for consideration of depth of cover used in the USA is also plotted in the same figure by a straight line. Initially, it shows a very high stress compared to the actual one but at the end it becomes quite low compared to the actual one. Hence, the weighted approach for the USA may be adopted where the seam is flat and highwall is very steep i.e. about 75–80° but in an Indian scenario it does not take into account the complexities associated with a highwall mining site.

When the same study is extended for top seam a different pattern of increase or decrease in stress was observed (Figure 3.3). The initial steep rise in vertical stress

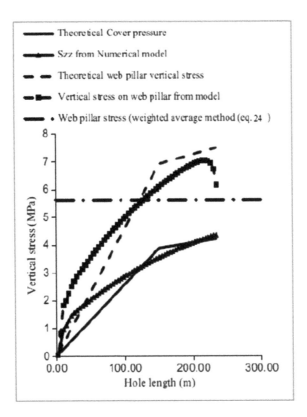

Figure 3.3 Comparison of theoretical and numerical vertical stress acting on seam horizon before extraction and on web pillar after extraction for top seam (Porathur et al., 2013).

near the highwall face is not found here but the observed vertical stress is definitely quite high, mainly due to induced stresses from open-pit excavation. However, a drop in stresses at the end of cut is observed as with the previous study. Hence, it can be said that the initial steep rise in the case of bottom seam was due to toe of slope.

Another issue that was addressed is when the seam is up-dipping at about 5°. The same model is extended for up-dip seams and the trend obtained is shown in Figure 3.4. Here the maximum vertical stress was observed at the highwall face which drops down and rises again for some distance and then decreases gradually with increasing depth of penetration as the depth of cover is decreasing. In this case, the end effect of the web cut reaching solid boundary was found to be insignificant and therefore the vertical stresses over web pillars were not studied again. Consequent to the study above, it was concluded that the highwall is subjected to asymmetrical loading induced by open-pit excavations, damage induced by blasting, variation in seam depth and gradient, and the end effect.

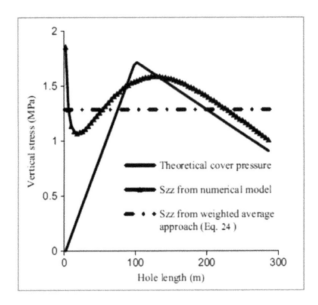

Figure 3.4 Comparison of theoretical and numerical vertical stress acting on seam horizon for bottom seam at up gradient of 5° (Porathur et al., 2013).

Hence it is recommended that Szz_{max} obtained from site-specific numerical modelling in estimation of maximum stress over web pillars is used, which is obtained as:

$$P_{max} = \frac{Szz_{max}(W + W_c)}{W} \tag{3.13}$$

where P_{max} is the maximum vertical stress on web pillar (MPa), and Szz_{max} is the maximum vertical stress acting on the seam, may be evaluated using numerical modelling (Porathur et al., 2013). The design methodology explained in Chapter 4 was focused on numerical modelling to determine Szz_{max}, instead of considering the theoretical maximum depth of cover or a weighted average depth.

3.2.8 Safety factor

Safety factor of the pillars is calculated using the following equations:

$$\text{Safety Factor, } S.F. = \frac{Strength\ of\ web\ pillar}{Load\ on\ web\ pillar} \tag{3.14}$$

On the basis of past experiences from Indian coalfields it has been observed that a pillar safety factor of more than 2.0 is stable long-term, i.e. for many decades. A safety factor between 1.5–2.0 may be taken as stable medium-term, i.e. stable for a few years, and if the safety factor of the pillar is 1.0, it may be treated as stable short-term, with a standup time of a few weeks or a month. CSIR-CIMFR has framed design and development norms for highwall mining in India (Porathur et al., 2013) and in this regard a circular has been issued by the Directorate General of Mines Safety, India (DGMS Technical Circular No. 06 of 2013). A detailed discussion on safety factors at all the highwall mining sites in India was elaborately put forward in Chapter 4. To ensure the safety of important surface properties over highwall extraction panels, the web pillars were designed considering long-term stability in most of the case examples cited in Chapter 4.

3.3 EQUIVALENT WIDTH CONCEPT

Highwall web pillar width needs to be estimated based on experiential, numerical, empirical-mechanistic and hybrid modelling. The web pillars in highwall mining are long, narrow and mostly slender, contrary to the square and rectangular shaped coal pillars in underground coal mines. As the length is much higher compared to the width, it can have a two-dimensional plane-strain effect. Most of the above empirical pillar strength equations for highwall mining web pillars are adaptations from rectangular pillar strength formulations. It has been found that direct use of rectangular pillar strength equation to web pillars results in an underestimation of its strength. To account for the shape effect of the long web pillars, an equivalent width concept has been proposed by several researchers to make empirical pillar strength equations usable for highwall mining web pillars.

Wagner suggested the equivalent width concept for non-square pillars and for very long pillars as $W_e = 2W$. But based on highwall mining experience in Australia, $W_e = 1.5W$ is found to be more appropriate when workings are at shallow depth and height of seam is less than 3.5 m (Medhurst, 1999). Wagner (1974) conducted an extensive study on *in situ* coal pillars to understand the phenomena of failure in coal. It was found from his study that failure commences at the perimeter of a pillar and then progresses inward. It supported the concept established by Wilson (1973) that stress over pillar is non-uniform as shown in Figure 3.5 and confining stresses develop a pillar core in the centre.

The data of Wagner's study (1974) was back-analysed and a more appropriate representative data for strain-softening model for coal simulation was prepared by

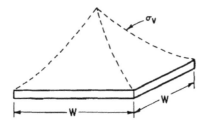

Figure 3.5 Non-uniform stress gradient over a pillar (Wilson, 1973).

Robert et al. (2005) so as to develop a suitable equation for strength estimation of long slender pillars. This model is fitted to Salamon and Munro's pillar design equation (1967) which is commonly used in Australia for coal pillars' strength estimation in cases other than highwall mining. Two-dimensional models of rib pillars are prepared and strain-softening is introduced through a developed model. Accordingly, two equations are proposed as given below; named as upper-bound equation and lower-bound equation. For normal conditions the upper-bound equation (Equation 3.15) is suggested to be more suitable, while a lower-bound equation (Equation 3.16) is considered suitable for mining conditions where the roof and floor are weak (Robert et al., 2005).

$$S_{max} = 10.95 \left(\frac{W_{min}}{h} \right)^{0.46} \quad upper\text{-}bound \tag{3.15}$$

$$S_{min} = 9.75 \left(\frac{W_{min}}{h} \right)^{0.29} \quad lower\text{-}bound \tag{3.16}$$

where
S_{max} – pillar strength from upper-bound equation, MPa
S_{min} – pillar strength from lower-bound equation, MPa
W_{min} – width of rib pillar or minimum width of long pillar, m
h – pillar height, m

These formulae are applicable for W/h ratio ranging from 0.5–2.5. Although the model used was for a case study data and design was found to be reasonable, in the absence of an empirical database it may not give any idea regarding failure probability of long pillar.

For Indian highwall mining cases, the CSIR-CIMFR equation has been used with good success so far. Although Wagner's concept of equivalent width $W_e = 2W$ is made use in the CSIR-CIMFR equation, it was found from extensive numerical modelling studies that strength estimated using the above equivalent width concept is good enough for web pillar with a slenderness (W/h) ratio greater than 1.0. For very slender web pillars with $W/h < 1.0$, the strength estimated needs to be reduced by up to 25% to give more accurate results.

3.4 COMPARISON OF PILLAR STRENGTH ESTIMATION APPROACHES

To analyse the various approaches of estimating pillar strength, a comparison is made for better understanding of the pillar design. For comparison the Mark-Bieniawski equation is considered as base because it has been used successfully for design in many USA highwall mining sites. The working depth is taken as 52 m, height of seam as 2 m, and width varied to obtain different W/h ratios. Apart from the basic approaches discussed in Sections 3.2 and 3.3, a few other observations need to be mentioned.

Ryder and Ozbay (1990) suggested a multiplying constant of 1.3 to strength predicted by square pillar equation for long pillar strength estimation. Hence strength obtained by the CSIR-CIMFR equation for square pillar is multiplied by 1.3 to

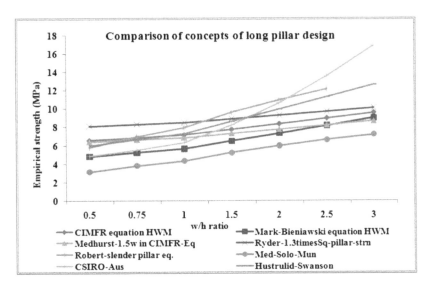

Figure 3.6 Comparison of various concepts of long pillar design (Verma et al., 2014).

plot the strength based on Ryder and Ozbay's observation (1990). To consider the Medhurst observation of effective width, it is taken as $1.5\,W$ in CSIR-CIMFR equation and plotted as 'Medhurst-$1.5\,W$ in CSIR-CIMFR-Eq' in Figure 3.6. Robert et al. (2005), developed the slender pillar equation for normal conditions (Equation 3.15), which is plotted up to $W/h = 2.5$ since it is applicable for maximum W/h of 2.5. All the computed results are presented in Figure 3.6.

From Figure 3.6, it can be seen that the slender pillar equation of Robert et al. (2005) predicts pillar strength on the higher side especially beyond $W/h > 1.0$. The trend of all other concepts is more or less the same. Medhurst's $W_e = 1.5\,W$ in CIMFR-Eq. seems to be practical, but beyond $W/h = 2.5$ it suggests stagnation in strength. Otherwise up to $W/h = 2.5$ difference in predicted strength is insignificant with the CSIR-CIMFR equation whether it is $1.5\,W$ or $2\,W$. The multiplying constant suggested by Ryder and Ozbay (1990) gives higher value of predicted strength while Medhurst $W_e = 1.5\,W$ in the Salamon-Munro equation predicts strength on the lower side. In addition to that it has the limitation of applicability to a shallow working depth and seam height less than $3.5\,m$. In India, where the working depth of highwall mining extractions is quite high, the CSIR-CIMFR equation with $W_e = 2\,W$ and the Mark-Bieniawski equation for web pillar match quite well, though the strength predicted by the CSIR-CIMFR equation is a little higher. In view of the limited field experience in Indian highwall mining, and above all the steeply dipping seam with 10–$14°$ inclination at a high working depth with the existence of multiple-seam in close proximity, it is better to adopt the CSIR-CIMFR equation with '$2\,W$'' as effective width even though it predicts a little higher strength than the Mark-Bieniawski equation. However, with more on-site experience this can be further refined in the near future. The comparison of strengths predicted by the Mark-Bieniawski, CSIR-CIMFR and CSIRO equations for Seams I and II in panel 'A' and panel 'B' of an Indian case example is presented

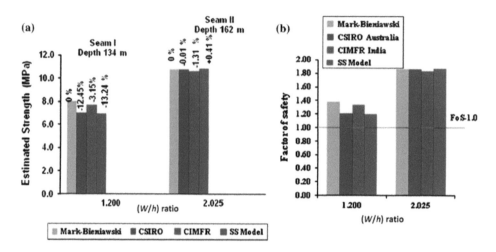

Figure 3.7 Estimated strength versus *W/h* ratio in panel A (OCP-II) of end-user site (Verma et al., 2014).

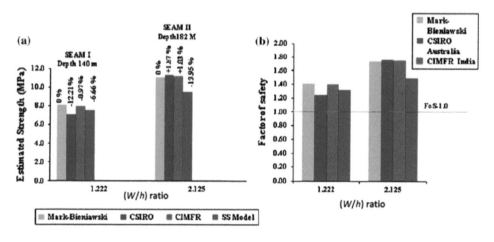

Figure 3.8 Estimated strength versus *W/h* ratio in panel B (OCP-II) at end-user site (Verma et al., 2014).

in Figures 3.7 and 3.8 (CSIR-CIMFR, 2008). This case example is discussed in detail in Chapter 4.

3.5 SPAN STABILITY

Several rock-geologic parameters affect the span stability in mining excavations. Based on past experience, there have been good empirical classification systems such as RMR, Q-system (Q) and Coal Mine Roof Rating (CMRR) that could be used for designing the span. In Indian conditions, span and support are designed based on RMR system.

CSIRO (Shen and Duncan Fama, 1996) established a methodology for estimating roof span based on Laminated Span Failure Model (LSFM).

There are four commonly used methods viz. RMR system (Bieniawski, 1976) and Modified Rock Mass Rating for underground coal mines by Indian researchers (Venkateshwarlu et al., 1989), Q-system (Barton et al., 1974) and the CMRR system (Molinda and Mark, 1994). The advantages of such empirical systems are noted below.

The exploration process routinely obtains both geologic and geotechnical data such as the following parameters:

- Rock Quality Designation
- Fracture Frequency Index (number of fractures per metre)
- Joint Roughness Coefficient
- Point Load Test (axial and diametral)
- Fracture formations and spacing are obtained during the exploration process and mapping of the highwalls

Since the information obtained is the same data required for use in one of several RMR or Roof Mass Rating systems, these systems can readily be used. The results may then be used to focus on the areas where roof conditions are adverse. A CMRR system (Molinda and Mark, 1994) and an RMR (Bieniawski, 1989) are obtained from each borehole. The CMRR is designed for coal measure strata, and the input data required addresses the geotechnical characteristics, therefore the results more closely address the roof stability. The CMRR normally uses the bolted height as the target area. Since highwall mining does not have a supported roof, a suitable target depth can be assumed.

For Indian conditions, web pillars of higher FoS are recommended to eliminate any failure of the rock mass and to avoid any eventual surface subsidence. One of the disadvantages of having a highwall mining panel with lower FoS coupled with barrier pillar of higher FoS is that if the roof strata is weak above two metres into the roof, major roof stability problems can be expected and the ability to mine the overlying reserve needs to be closely evaluated. This situation can be avoided if a series of highwall excavations with web pillars of higher FoS are designed without barrier pillars.

The relative values in rock classification systems–such as CMRR and RMR–obtained through exploration holes can be used to identify hazardous sites, which will aid in focusing on potential problem areas.

3.5.1 Laminated Span Failure Model (LSFM)

The LSFM for estimating roof span was developed by CSIRO, Australia. The LSFM models the roof as a series of thin rock beams which are detached from each other and which can progressively fall out of the roof. The LSFM combines an equation to calculate the maximum stable span with a simple technique to estimate the likelihood of span failure and the height of fall for a given highwall mining span. A detailed analysis on span stability citing different case examples is given in Chapter 6.

In two dimensions, the LSFM considers the immediate roof layer as a thin and long beam which is detached from the overlying rock. Cracks can develop near the abutments and at the mid-span as the beam deflects under its self-weight. The final failure of the immediate roof beam is caused by snap-through at the mid-span.

According to the LSFM, the maximum stable span is controlled by the roof layer thickness, Young's modulus and compressive strength, and load applied to the layer. The maximum stable span is calculated by using a simple closed-form solution as follows:

$$S = 1.873 \cdot t \cdot \left(\frac{E}{p}\right)^{1/4} \tag{3.17}$$

S signifies the span (m); t is the roof layer thickness (m); E is Young's modulus (MPa) and p is the distributed vertical load on the roof layer (MPa), in most cases being the self-weight of the layer. But it can be greater due to additional load from the overlying layers, particularly when the overlying layers are not horizontal.

Similarly, there are two other models developed by CSIRO (Shen and Duncan Fama, 1996) for estimating span for coal-bearing rock strata viz. laminated roof-cracking model (LRCM) and laminated roof-shearing model (LRSM). LRCM is designed to predict tensile cracking of the roof. It considers the immediate roof after bedding planes have separated. The beam is considered to be loaded under its own weight and *in situ* horizontal stress. The final form of the equation is given below:

$$S = t\sqrt{\left(\frac{210E(\sigma_h + \sigma_t)}{p^2}\right)} \left[(1 + \sqrt{1+k})^{1/3} + (1 - \sqrt{1+k})^{1/3}\right] \tag{3.18}$$

$$k = \frac{35pE}{36(\sigma_h + \sigma_t)^2} \tag{3.19}$$

p is taken as the self-weight or the unit weight of the beam; σ_h and σ_t being minor horizontal stress and tensile strain respectively.

The third model is the LRSM which is based on the assumption that after cracking near the abutment, the beam loses all or part of its cohesion at its ends, depending on the length of the cracks (Shen and Duncan Fama, 1996). It the horizontal confinement is low, shearing is assumed to occur along the cracked surfaces and the beam may slide down because of complete shear failure near the abutment. With this LRSM, it is considered that the whole thickness of the beam has been cracked and no bending moment exists at the abutments. The model results in a non-linear equation of the form given below:

$$\frac{17}{280}\frac{p^2 \tan\phi}{E}\left(\frac{S}{t}\right)^6 - \frac{p \tan\phi}{2}\left(\frac{S}{t}\right)^2 + \frac{p}{2}\left(\frac{S}{t}\right) - \sigma_h \tan\phi = 0 \tag{3.20}$$

This equation is solved for span, S.
where,
S – span (m)
t – roof layer thickness (m)
ϕ – friction angle of coal/roof interface
E – Young's modulus (MPa)

p – distributed vertical load on the roof layer (MPa), in most cases being the self-weight of the layer. But it can be greater due to additional load from the overlying layers, particular when the overlying layers are not horizontal.
σ_h – horizontal stress (MPa).

Comparative results show that both LRSM and LSFM predict very similar maximum stable spans. These models can be used for a preliminary estimate of roof span. A detail assessment of span stability is given in Section 6.6.2 of Chapter 6.

However, there are certain drawbacks in such analytical models viz.:

- They are idealised models with assumptions and do not consider the effects of geometric non-linearity and material non-linear parameters, do not incorporate appropriate failure criterion, and are not derived on the basis of field experiences;
- LSFM does not consider the tensile strength of the roof layer and horizontal *in situ* stresses, while LRCM does not consider the cohesion and friction properties, and LRSM does not consider the strength properties;
- They do not consider the composite effect of the beam;
- They do not explicitly consider fracture mechanics properties;
- In reality, the roof span is affected by many other parameters viz. jointing, water conditions, slake durability, etc.

Hence it is suggested to initially evaluate the span stability based on RMR, Q and CMRR approaches. Further numerical modelling needs to be undertaken to reassess the span stability and ensure that the models are well calibrated against stress and displacement which are obtained through deploying stress-cell or extensometers by drilling boreholes from the surface, or having low-angled drilling if dump prevents drilling vertical holes. It is recommended to have these in the pillars and in the excavations to monitor pillar stresses and roof deformations.

3.5.2 Simplified method based on roof layer thickness

CSIRO, Australia, has also developed a simple method to assess the entry span stability based on the thickness of roof layers for laminate roof types (Shen and Duncan Fama, 2001). A simple formula was derived to calculate the maximum stable unsupported span, S.

$$S = 2 \cdot \left[\frac{4}{3\sqrt{3}} \right]^{1/4} \cdot t \cdot \left(\frac{E}{p} \right)^{1/4} \tag{3.21}$$

where, t – thickness of the roof plate (distance between parting planes)
E – Young's Modulus of the roof rock
p – self-weight of the plate

This formula, with some refinements relating to possible crushing or yield of the rock, has been used for ongoing assessment of pits being highwall mined at the Moura Mine, Australia, and appears to give reasonable predictions of the unsupported span performance.

3.5.3 Span stability based on Rock Mass Rating (RMR)

RMR is a geo-mechanical classification system for rocks, developed in 1973 in South Africa by Professor ZT Bieniawski. It was later modified for Indian conditions by Venkateshwarlu et al. (1989) and the classification system was called CMRI-RMR, 1989.

A number of empirical design approaches have been developed for Indian conditions which have been in use for decades in India for roof support. Indians have built up confidence in using RMR for coal mines for support design. This data is available in most of the coal mines in India or it can be generated with local expertise in India.

Modified RMR has been used greatly in India for Indian conditions. The modified RMR is called CMRI-RMR (CSIR-CIMFR, 1987; Venkateshwarlu et al., 1989) which depends on the following parameters:

- Layer thickness – determining layer thickness is important as it is one of the important parameters in the classification system. For a sandstone roof, it is determined based on frequency of bedding planes per metre; for shale rooves it is determined based on the frequency of prominent lamination, and for the coal roof, it is determined as frequency of prominent bands.
- Structural features include major faults, slips, joints and other sedimentary features, such as sandstone channel intrusion, plant impression, etc. CMRI (presently renamed as CSIR-CIMFR) has developed an approach to quantify the geological features based on nature and magnitude of influence of different features which is based on their local expertise and experience (Venkateshwarlu et al. 1989).
- Weatherability – measured by Slake Durability Index (SDI) in the 1st cycle as per ISRM or BIS/IS-1981 standards (#10050/1981 BIS standards, New Delhi).
- Compressive strength – determined in the laboratory based on Indian BIS/IS-1979 standards and can be determined in the field using point load tester on irregular samples or by Schmidt hammer.
- Groundwater condition – based on the seepage of groundwater and expressed as ml/min.
- The CMRI-RMR rating is also adjusted for various geo-mining conditions, such as depth, later stress, influence of adjacent and overlying working and mode of excavation of galleries.

Empirical designs based on the classification systems have limitations, but applied appropriately and with care they are valuable tools. Therefore, for Indian conditions, it is suggested that empirical design charts such as RMR are used as first pass design with awareness of its limitations. It is further important to supplement empirical design charts with calibrated numerical modelling with properties appropriate to the site conditions and incorporating the roof and floor seam conditions and any geological features such as faults, slips and groundwater.

Safe standup time of the roof span can also be evaluated using RMR and the roof span as given in Figure 3.9. For highwall mining scenarios, a web cut generally has a maximum standup time of a couple of days.

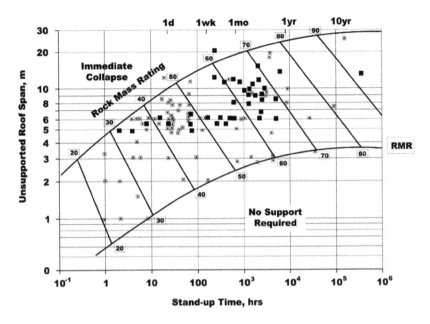

Figure 3.9 Evaluation of safe standup time using RMR (Barton and Bieniawski, 2008).

3.5.4 Span stability based on CMRR

The CMRR was developed by the US Bureau of Mines as an engineering tool to quantify descriptive geologic information for use in coal mine design and roof support selection (Molinda and Mark, 1994). Studies conducted in underground development headings have shown that the CMRR can be used to evaluate unsupported span stability (Mark, 1999). The CMRR system combines the results of many years of geologic ground control research with worldwide experience in rock mass classification systems. Like other classification systems, the CMRR begins with the premise that the structural competence of mine roof rock is determined primarily by the discontinuities that weaken the rock fabric. The CMRR makes four unique contributions: (1) it is specifically designed for bedded coal measure rocks, (2) it concentrates on the bolted interval and its ability to form a stable mine structure, (3) it is applicable to all US coal seams, and (4) it provides a methodology for data collection. The CMRR weights the geotechnical factors that contribute to roof competence, such as bedding, spacing, jointing and groundwater, and combines them into a single rating on a scale of 0–100. Using only simple field tests and observations, the CMRR can be calculated from roof falls, overcasts and highwall exposures. Over time, CMRR uses ARMPS-HWM developed by the National Institute for Occupational Safety and Health (NIOSH), USA, which incorporates modifications for designing web and barrier pillars for highwall mining.

3.5.5 Span stability based on Q-system

Q-system was developed for classification of rock masses and was first introduced in 1974 by the Norwegian Geotechnical Institute (Barton et al., 1974). It has been widely

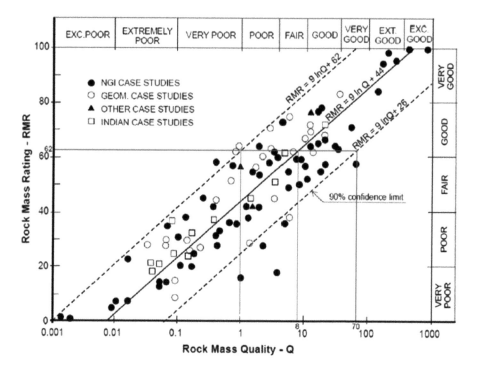

Figure 3.10 RMR and Q correlation chart (Bieniawski, 1984 c/f Palmstrom, 2006).

used as a classification system in rock engineering and design. Many improvements and/or adjustments of the system have been published over time. The input parameters to Q are: Rock Quality Designation (RQD), Joint Set number (J_n), Joint Roughness number (J_r), Joint Alteration number (J_a), Joint Water reduction factor (J_w) and Stress Reduction Factor (SRF). These are grouped below:

- Relative block size (RQD/J_n)
- Inter-block shear strength (J_r/J_a)
- Active stresses (J_w/SRF)

Using the above parameters, the ground quality Q (Barton et al., 1974) is defined as:

$$Q = (RQD/J_n) \times (J_r/J_a) \times (J_w/SRF) \tag{3.22}$$

There exists a correlation between the RMR and the Q as shown in Figure 3.10 (Bieniawski, 1984; Jethwa, 1981). The upper and lower bounds are specified in the figure with 90% confidence equations. On an average basis, RMR and Q are related by the following equation:

$$RMR = 9 \ln Q + 44 \tag{3.23}$$

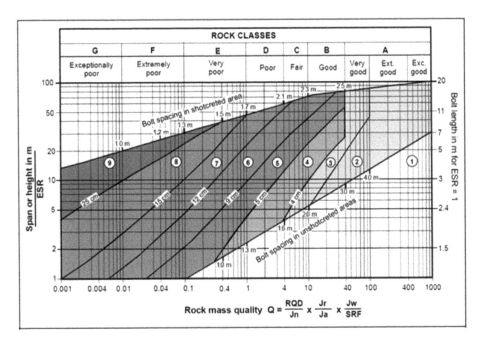

Figure 3.11 The Q support chart (Grimstad and Barton, 1993).

Once *RMR* is determined, *Q* can be estimated or vice versa. However, it should be used with caution.

Once the Q values are determined, the unsupported span can be estimated by Figure 3.11 (Grimstad and Barton, 1993). It should be noted that the design chart given in Figure 3.11 is used for tunnelling purposes, however, for coal mines it is subjected to verification and validation.

As a case example of the Ramagundam area of M/s Singareni Collieries Company Ltd. (SCCL), *Q* is 9.2 and assuming the excavation support ratio of 1.6 for highwall mining span of 3.6 m, then the 'equivalent dimension', D_e of the excavation – which is the ratio of span to the *ESR* – is evaluated as 2.25. For such D_e and *Q*, a span of 3.6 m is stable as the point (9.2, 2.25) falls in the unsupported region – 1 (Figure 3.11). It is, however, suggested to acquire more *RMR* and *Q* data while estimating the support zone for highwall mining excavations and update these classification charts, at least for the first pass analysis.

3.6 APPLICATION OF NUMERICAL MODELLING TOOLS

Various numerical modelling software viz. Fast Lagrangian Analysis of Continua in two dimensions (*FLAC*) and in three dimensions (*FLAC3D*); Universal Distinct Element Code (*UDEC*), Three-dimensional Distinct Element Code (*3DEC*); two-dimensional Finite-Element Analysis Software Package *FESOFT*; three-dimensional stress analysis program (*3STRESS*); boundary element programme incorporating laminated medium

(*LAMODEL*) and a two-dimensional finite-element code (*AFENA*) are available for simulating different rock types under different mining conditions. Either a single parameter or range of parameters can be analysed using them.

FLAC and *UDEC* are extensively used for grid creation, application of boundary conditions, selection of appropriate constitutive (material) models, solution of the static and dynamic equilibrium states, simulation of stress states during excavation, installation of structural support and stability analysis. The built-in programming language called *FISH* is also used to manoeuvre the *FLAC* model.

Extensive usage of numerical modelling for applicability, design and safety of various highwall mining sites in India and Australia are elaborated in Chapters 4 and 6. Site-specific applications of *FLAC* and *UDEC* are presented below (Verma et al., 2014).

3.6.1   Fast Lagrangian analysis of continua: *FLAC* and *FLAC3D*

Fast Lagrangian analysis of continua is a commonly used code for stability and other analyses of various rock engineering problems. *FLAC* was used for elasto-plastic analysis to verify pillar strength under different circumstances arising from the extraction strategy planned at the highwall mining sites of Ramagundam Opencast Project-II and Medapalli Opencast Project of M/s. Singareni Collieries Company Ltd. (SCCL); Quarry SEB and AB, West Bokaro of M/s. Tata Steel Ltd. (TSL) and Sharda Opencast Project of South Eastern Coalfields Ltd. (CSIR-CIMFR, 2008; 2009; 2011 and 2014). It was also used for FoS analysis of individual seams and to visualise multi-seam interactions (Loui et al., 2008). In the United States there are several cases where highwall miners are exploiting previously augered highwalls and by using *FLAC* models the strength of a highwall web pillar containing auger holes was obtained (Zipf, 2005).

3.6.2   Distinct element code: *UDEC* and *3DEC*

In Australia, *UDEC* was used to verify the findings of the Coal Roof Failure Model (CRFM). *UDEC* results from back analysis of a failure in the central Moura Mine supported the findings of the Local Mine Stiffness (LMS) theory (Adhikary et al., 2002). Similarly, the cause of a failure in Oaky Creek mine, Australia, as predicted by *UDEC* and the LSFM, was in agreement with data from field investigations (Shen and Duncan Fama, 1999). In the Colowyo mine in northwest Colorado, USA, *UDEC* was used to visualise multi-seam interactions and the corresponding stability scenario in the seam roof and floor (Molinda and Mark, 1994). *UDEC* was also used to analyse a failure in Yarrabee mine, Australia (Duncan Fama et al., 2001). Good correlations were obtained between micro-seismic monitoring data and *3DEC* results of analysis of the rock mass response as a result of a highwall mining operation in Pit 20D Upper North, Moura Mine (Shen et al., 2000).

Chapter 4

Indian case studies

4.1 INTRODUCTION

India's domestic energy requirements are met by coal combustion to the extent of about 60%. Currently the country produces about 620 MT of coal per year, about 88% of which comes from opencast mines. There has been a great need to substantially increase coal production in the coming decade in order to meet the energy needs of the growing economy of the country.

Unfortunately, many of the Indian opencast mines are reaching their pit limits. The existence of surface dwellings in many places limits the expansion of the currently operational opencast mines. Also, in many cases, the overburden becomes so high that coal extraction becomes uneconomical. However with the use of highwall machines, wherein a cutter is placed on the top of a continuous miner and taken through a conveyor inside the seam, which can be almost 250–500 m deep inside, a sizeable percentage of such locked-up coal can be extracted. Highwall mining technology is a modified version of earlier auger mining. It facilitates the recovery of locked-up deposit in open-pit highwalls, below villages, road, and other surface structures. Mining operations consist of driving series of parallel roadways at intervals in a planned way to recover considerable amounts of coal from the highwall. As already elaborated, the method offers great safety to mining personnel because of remotely-controlled highly-mechanised mining operations and unmanned extraction drivages. Highwall mining can be successful only if a feasible design of extraction is made, taking into consideration the complex geology and existence of multiple coal seams in Indian coalfields, which is unique in many respects.

In India, the following mining sites were considered for highwall mining operations: (i) Ramagundam Opencast Project-II of M/s Singareni Collieries Company Ltd. (SCCL), (ii) Medapalli Opencast Project of M/s Singareni Collieries Company Ltd. (SCCL), (iii) Quarry SEB and AB, West Bokaro of M/s Tata Steel Ltd. (TSL) and (iv) Sharda Opencast Project of M/s South Eastern Coalfields Ltd. (SECL). CSIR-CIMFR designed the extraction methodology of highwall mining of the projects mentioned in serial nos. (i), (ii) and (iii) of which the first highwall mining system in India started on 10 December 2010 at Opencast Project-II (OCP-II), Ramagundam Area-III (RG-III) of Singareni Collieries Company Ltd. (SCCL). Subsequently, in February 2011, M/s Cuprum Bagrodia Limited started highwall mining at Sharda Opencast Project of South Eastern Coalfields Ltd. (SECL) by cutting a trench for operation of highwall miners. CSIR-CIMFR's association in that project was in the areas of stability analysis of highwall slopes and pre-split blasting. All these case studies are elaborated in the following sections serially.

4.2   SITE A

HIGHWALL MINING AT RAMAGUNDAM OPENCAST PROJECT-II (OCP-II) OF M/S SINGARENI COLLIERIES COMPANY LIMITED (SCCL)

In Ramagundam OCP-II, seams IV, III, IIIA & II were targeted for highwall mining in a small patch of approximately 750 m to 1 km length against GDK 10A underground mine with sufficient barrier between them (Table 4.1). Coal grades varied from C to E and seams were rated with degree-1 gassiness (Tables 4.2 and 4.3). The general dip of

Table 4.1   Seam details.

Seam	Depth of cover (m)	Thickness (m)	Parting (m)
I	95–154	5.2–6.7	–
II	112–182	3.1–4.0	21–29 with seam I
IIIA	155–229	0.9–1.8	31–47 with seam II
III	184–255	11–12.2	24–28 with seam IIIA
IV	199–270	3.7–4.1	4–5.6 with seam III

Table 4.2   Degree and classification of gassiness.

Degree of gassiness of coal seam	Classification of gassiness
I	<0.1% of inflammable gas in the general body of air and rate of emission of such gas is less than 1 m^3/t of coal production
II	>0.1% of inflammable gas in the general body of air and rate of emission of such gas is greater than 1 m^3/t but less than 10 m^3/t of coal production
III	Rate of emission of the inflammable gas is greater than 10 m^3/t of coal production

Table 4.3   Calorific values of non-coking coal.

Grades	Calorific value (in kcal/kg)
A	>6200
B	5601–6200
C	4941–5600
D	4201–4940
E	3361–4200
F	2401–3360
G	1301–2400

the coal seams was 1 in 4.2 and the maximum length of entries would be 350 m from the highwall. The estimated coal recovery would be over one and half million tonnes with a recovery rate of around 35 to 40%. As there were five seams lying over each other with varying parting thicknesses, a feasibility report on the design and safety aspects of highwall mining was carried out based on the following issues:

1 Geo-technical studies for web and barrier pillar design for all the five workable seams on the basis of

   a) long-term stability of seams IV, III, IIIA & II in ascending order,
   b) short-term stability of the top-most seam I, and
   c) overall long-term stability of seam III and seam IV workings, since they were contiguous (in close proximity).

2 Layout of highwall mining web cuts for all the five seams based on the above design. The need for barrier pillars for seams II to IV.
3 Design of controlled blasting parameters for the open-pit bench blasting to reduce ground vibration and flyrock up to a distance of 100 m from the highwall mining site.
4 Study of highwall slope's stability.

4.2.1 Extraction strategy and pillar design

At OCP-II of SCCL, M/s Advanced Mining Technology Private Limited (AMT) had been contracted to extract the remnant coal within the mine boundary using the continuous highwall mining method. The highwall extractions were planned with ADDCAR Highwall Mining Systems, USA. This was the first highwall mining operation in the country, in fact it was the first continuous highwall mining operation in Asia. The available seams were seam IV, seam III, seam IIIA, seam IIIB, seam II and seam I, in ascending order. Out of the above six seams, seam IIIB was declared unworkable due to insufficient thickness. Hence five seams were targeted for highwall mining: seam IV, seam III, seam IIIA, seam II and seam I. The major coal reserve was locked up in seam IV and seam III occurring in close proximity separated by a parting of 4–5 m of sandstone. At places this parting was found reduced to less than 3 m, and at a very few locations seam III and seam IV occurred as one. The general dip of all the seams was 1 in 4.5 (12.5°).

About a 900 m length of the highwall at the pit bottom was exposed and was available for highwall mining operations. The extent of the highwall mining block was defined taking into account a minimum of 20 m barrier from the OCP-II boundary. The highwall had an overall slope angle of about 47°. Therefore, the bottom most seam IV had the highest mineable reserve and the top-most seam I had the least. The mining block, with two extraction panels A and B is as shown in Figure 4.1.

The major points considered from a geo-technical perspective for the highwall mining operations of all the five workable seams are as follows:

1 A suitable sequence of extraction of the seams, ensuring maximum extraction from all the workable seams and minimum damage to the overlying seams

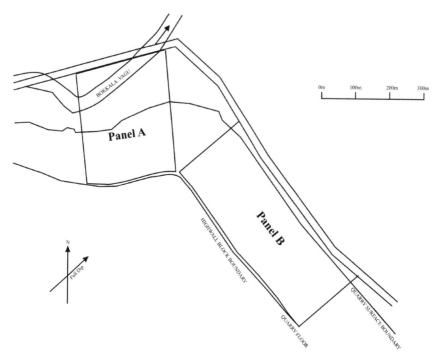

Figure 4.1 Plan showing the proposed highwall mining block at OCP-II (Porathur et al., 2013).

2   An optimum extraction strategy for each seam with a combined strategy for seam III and seam VI
3   Design of web pillars and barrier pillars, if necessary, for each panel in all the workable seams, ensuring optimum recovery and protection of upper seams
4   Protection of surface features, if any, and the proposed "punch entries" for the underground mine

The above geo-technical aspects had been dealt with in an empirical pillar design approach for the mining block supported by numerical modelling studies using *FLAC3D* software by the ITASCA Consultancy Groups, USA.

4.2.2   Geo-technical data

Pertinent geo-technical data for the design of OCP-II highwall mining had been obtained from SCCL and AMT. A representative borehole section showing all the workable seams and the lithology is given in Figure 4.2. Depth of cover, seam thickness and the parting thickness of all the workable seams are given in Table 4.1.

The physico-mechanical properties of the seams and their respective immediate roof are given in Table 4.4. These properties were tested earlier by the National Institute of Rock Mechanics, India (NIRM). The values given in Table 4.4 were used for the current design.

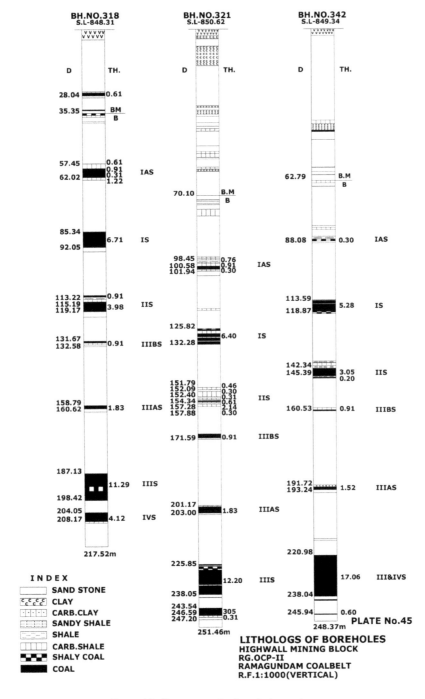

Figure 4.2 Representative borehole sections.

Table 4.4 Physico-mechanical properties of the seams and immediate roof.

Strata	σ_c (kg/cm^2)	σ_t (kg/cm^2)	E (kg/cm$^2 \times 10^5$)	ν
Seam I immediate roof	113	9	0.50	0.50
Seam I	365	24.80	0.274	0.285
Seam II immediate roof	–	–	–	–
Seam II	340	24.27	0.26	0.26
Seam IIIA immediate roof	366	–	1.01	0.36
Seam IIIA	395	22	0.35	0.31
Seam III immediate roof	251	–	1.01	0.42
Seam III	415	27	0.32	0.34
Seam IV immediate roof	315	–	1.07	0.43
Seam IV	414	23	0.34	0.33

Table 4.5 Rock mass properties of immediate roof.

Parameter	Seam I		Seam II		Seam IIIA		Seam III		Seam IV	
	Low	High	Low	High	Low	High	Low	High	Low	High
RQD (%)	40	100	75	97	60	80	48	52	48	52
RMR	33	74	40	74	38	79	33	70	33	65
Q	0.29	28	0.64	28	0.51	49	0.29	18	0.29	10

The rock mass parameters used for strength reduction in the numerical modelling studies are as given in Table 4.5.

4.2.3 Strategy for extraction

The final pit slope had already been reached, indicating the end of all the possible opencast mining operations; the pit limit on the surface had also reached the limit of the mine, exposing all the seams as outcrops on the final pit slope. Further benching was not possible in the 47° final pit slope and therefore highwall mining could only progress upward from the bottom seam IV to the top-most seam I. After the extraction of seam IV it would be necessary to fill the pit bottom until the horizon of seam III, forming a platform for its extraction, and so on and so forth.

For extraction of the seams in ascending order, it was necessary that the left-out web pillars would ensure protection of the upper seams for the period that they were required. In this scenario, with an ascending order extraction sequence, a medium or long-term stability of the web pillars was required. The web pillars designed for medium or long-term stability should take care of the highest depth at which the pillars would exist. The end effect of the cuts at the boundary was also studied using numerical modelling to arrive at the maximum effective depth.

A seasonal stream Bokkala Vagu existed above the proposed mining block. The SCCL management had earlier decided upon shifting this stream outside the mining block boundary to avoid any significant surface features during highwall mining, however they were unable to do that. This could have allowed the design of top seam I with short-term stability of the web pillars given the absence of upper workable seams and surface features.

Since the seams III and IV were occurring in contiguity, a combined web pillar design was followed for the two seams, by maintaining verticality of extraction with pillar columnisation. The design in such cases would take care of the required safety factors for web pillars in both the seams, ensuring the possible recovery of coal from both seams.

The variation in seam thickness in seam III and seam IV was also an issue. As said earlier, the parting between the seams was also found to vary, which might go below 3 m in some places. During the ascending order extraction of these seams, the parting between the adjacent cuts should be sufficient enough to avoid its collapse and possible fall of the cutting system. This was of utmost importance while designing the 11–12 m thick seam III extractions vertically above the seam IV extractions. Owing to the greater thickness of seam III, the extraction might either be envisaged in two sections, bottom and top sections, or in one section by two-pass cutting.

As shown in Figure 4.1, the highwall mining block existed in an angular fashion, with a longer stretch approximately along the strike of the coal seams and a shorter length along the apparent dip (about 60° to the strike). Therefore, the highwall mining block was divided into panel A and panel B, for the portions along the apparent dip and strike, respectively. The web pillars were designed separately for these two panels taking into account the respective depth variations in order to ensure maximum recovery. A triangular portion of the coal reserve was left unmined between panel A and panel B. This was due to the practical difficulty for the ADDCAR highwall system to take fan cuts at a varying dip angle. However, this triangular portion acted as a barrier separating the two panels.

The extent of panel B on the southern side of the block was decided on the basis of the location of the punch entries for the underground mining. The punch entries were planned in the seam I horizon and the limit of the highwall mining block was decided taking into account the 45° angle of draw at the seam IV horizon.

4.2.4 Accuracy of web cuts

An Mk3 highwall mining guidance system, incorporating HORTA inertial navigation, was employed in the ADDCAR Highwall Mining Systems. On the basis of earlier mining data, the Mk3 system has an accuracy, in terms of the off-track deviation, of less than 0.2% of the penetration depth, i.e. a maximum of 20 cm of hole deviation for every 100 m of penetration depth.

This aspect was considered while designing the web and barrier pillars. The change in pillar thickness due to hole deviation resulted in a change of its strength resulting in a change in the safety factor. The expected web pillar thickness variation as per the above-mentioned accuracy levels was:

$$W_d = W \pm 2dL \tag{4.1}$$

where W_d is the deviated width, d is the percentage deviation and L is the maximum length of the extraction cut.

The above calculation accounts for the worst-case scenario of two adjacent holes deviating to their maximum towards each other.

The tolerance levels for safety factors were set as follows:

- For pillars designed for a minimum safety factor of 2.0, the tolerance level was kept as 10%. When the predicted deviation for the safety factor was greater than 10%, the web pillar size was readjusted to keep the lower limit within 10% of the designed safety factor.
- For pillars designed for a minimum safety factor of 1.5, the tolerance level was kept as 5%. The pillar size was redesigned as given above, whenever necessary.
- For pillars designed for short-term stability, a zero-level of tolerance was set, i.e. the web pillars were designed as per the lower limit of the safety factor ($W_d = W + 2dL$).

With the above discussed extraction strategy, empirical and numerical modelling studies were conducted to optimise web thicknesses for all the five workable seams.

4.2.5 Elastic numerical modelling for load estimation

The traditional tributary area load estimation with respect to the maximum depth had some shortfalls in the actual scenario. The maximum tributary area load estimated using the maximum depth of the workings might not be the maximum vertical stress acting on the web pillars in reality. There were three factors affecting the maximum vertical stress acting on the web pillar as given below:

a) the "end effect" at the maximum depth zone facing the solid boundary
b) effect of the highwall slope
c) the web and cut widths

Towards the end point of the extractions, the solid barrier facing the mine boundary had a tendency to reduce the vertical stress acting on the web pillar in its vicinity. Moreover, the highwall had a slope of about 47°, which reduced the vertical stresses to a considerable penetration depth. To find the maximum vertical stress likely to be acting on the web pillars it was prudent to use numerical modelling.

A three-dimensional elastic model had been constructed in *FLAC3D* taking two vertical planes of symmetry, one passing through the centre of the cut and the other passing through the centre of the web pillar. The pit slope excavation and the highwall extraction were made in sequence and the elastic model was run until an equilibrium was achieved. The elastic constants used as input parameters for the coal seam and overburden materials are given in Table 4.4.

Figure 4.3 shows the geometry of the model constructed for the numerical estimation of the load acting on the web pillars. Two models were run with 6 m and 22 m thick web pillars for finding the effect on web pillar width.

4.2.6 Results

Analysis of vertical stress acting on the web pillar revealed the following aspects:

1 The maximum vertical stress acted upon elements lying about 20 m from the boundary.

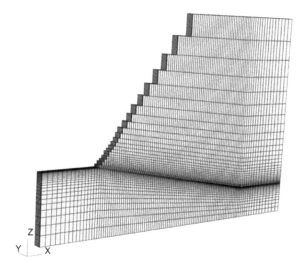

Figure 4.3 Elastic model geometry.

2   The maximum vertical stress magnitude was found to be roughly 90% of the maximum tributary area load corresponding to the maximum depth.

The above findings were used for the evaluation of the maximum load on the web pillars, used in the empirical calculation.

4.2.7   Empirical web pillar design

Web pillars were designed for seams III and IV combined, seam IIIA, seam II and seam I. The methodology adopted for the empirical design is as given below.

4.2.7.1   Estimation of pillar strength

The estimation of pillar strength was done using CMRI (presently CSIR-CIMFR) pillar strength formula:

$$S = 0.27\sigma_c h^{-0.36} + \left(\frac{H}{250} + 1\right)\left(\frac{W_e}{h} - 1\right) \text{ MPa} \tag{4.2}$$

where
S = strength of the pillar, MPa
σ_c = strength of 25 mm cube coal sample, MPa
h = working height, m
H = depth of cover, m
W_e = equivalent width of pillar, m
    = $2W$ for long pillar
W = width of web pillar, m

4.2.7.2   Pillar load estimation

The load on pillars was estimated using the Tributary Area Method, which reads as:

$$P = \frac{\gamma H (W + W_c)}{W} \tag{4.3}$$

where
P = load on web pillar, MPa
γ = unit rock pressure (0.025 MPa/m)
W_c = web cut width, m.

4.2.7.3   Safety factor

The safety factor of the pillars was calculated using the above equations as follows:

$$\text{Safety Factor, S.F.} = \frac{\text{Strength of web pillar}}{\text{Load on web pillar}} = \frac{S}{P} \tag{4.4}$$

The CMRI pillar strength equation was developed over a couple of decades after analysis of a large number of pillar stability observations from a gamut of Indian mining scenarios. On the basis of past experiences from Indian coalfields it had been observed that a pillar safety factor of more than 2.0 indicates long-term stability i.e. for many decades. A safety factor between 1.5 to 2.0 could be taken as medium-term stable, stable for a few years whereas a safety factor of 1.0 could be treated as short-term stable, with a standup time of a few weeks or a month.

4.2.8   Extraction of III and IV seams

As discussed earlier a combined strategy for seams III and IV was adopted due to their proximity (contiguity). These two seams form the base of the highwall block, and their stability was given high priority. The following points were kept in mind while designing web pillars for these two seams:

1   Pillars needed to be columnised in both the seams.
2   Seam IV was to be extracted to its full height of 4.0 m.
3   Seam III would require extraction in two passes to a total height of 8.0 m.
4   Positioning of the two-pass extraction in seam III was kept as high as possible after leaving 0.5–1.0 m coal in the seam III roof. This also maximised the parting between seams III and IV.
5   The web pillars would have long-term stability with a minimum safety factor of 2.0. Due to the wider variation in depth, alternate short and long-hole extraction patterns were designed to maximise the recovery. Barrier pillars within the panels were not required.

After several hit and miss attempts, the following web pillar widths were arrived at (Table 4.6).

In that table, L_L and W_L are the length of the long hole and the corresponding web pillar width respectively, and L_S and W_S are those for the short holes. This scheme of extraction by alternate short and long holes is shown in Figure 4.4.

Table 4.6 Web pillar sizes designed for safety factor 2.0 in seams III and IV.

| Seam (ht. ext., h) | Panel IV-A/Panel 3-A | | | | | Panel IV-B/Panel III-B | | | | |
| | Long holes | | Short holes | | | Long holes | | Short holes | | |
	Avg. length L_L	Web width W_L	Max. length L_S	Web width W_S	% ext	Avg. length L_L	Web width W_L	Max. length L_S	Web width W_S	% ext
IV (4 m)	345 m	18.7 m	115 m	7.6 m	21.0	225 m	20.5 m	135 m	8.5 m	23.3
III (8 m)	340 m	18.7 m	67 m	7.6 m	18.8	220 m	20.5 m	75 m	8.5 m	19.4

(a)

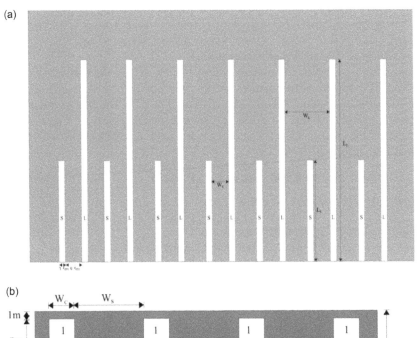

(b)

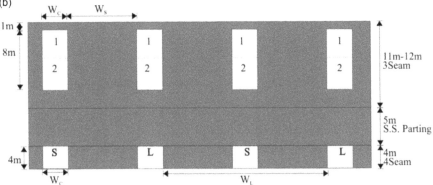

Figure 4.4 Schemes for extracting seams III and IV (a) plan (b) vertical section.

Table 4.7 Variation in safety factors with respect to the expected hole deviation for seams IV and III.

Seam	Panel IV-A/Panel III-A		Panel IV-B/Panel III-B	
	Long holes	Short holes	Long holes	Short holes
IV	±5.9%	±3.4%	±3.6%	±3.8%
III	±5.3%	±1.4%	±3.2%	±1.6%

The pattern of extraction as per the above design was like the one given in Figure 4.4. The long holes could be as long as possible within the boundary of the block. However, an average long-hole length was considered for extraction percentage calculations. The alternate short hole had to be restricted to the lengths as given in Table 4.7.

Variation in the safety factors due to hole deviation would be as given in Table 4.7.

The safety factor variations as given in Table 4.7 were within 10%, which was acceptable for the pillars designed for safety factors of 2.0, and therefore the design parameters given in Table 4.6 were acceptable.

As shown in Table 4.6, the overall coal recovery, as per the conservative safety factor of 2.0 in both the seams, amounted to only about 20.5%. Seams III and IV together contributed around 70% of the coal reserve from the block and therefore other options to improve the extraction percentage were explored. It had also been found that multiple seams in ascending order were extracted with web pillars designed for a 1.5 safety factor.

4.2.8.1 Alternate strategy for improving recovery

An alternate strategy adopted for improvement in recovery was as follows:

1 maintain columned pillars in seams III and IV
2 design web for a safety factor 2.0 in seam IV, which formed the base
3 design web pillar in seam III, maintaining verticality with seam IV, but with a 1.5 safety factor
4 perform an impact assessment on overall seams IIIA and II, had the seam III web pillars prematurely collapsed.

The design, which gave maximum recovery as per the above strategy, is given in Table 4.8.

The expected variations in safety factors due to hole deviation for this alternate configuration are given in Table 4.9.

The safety factor variations as given in Table 4.7 are within 10% for seam IV, which was okay for the pillars designed for a safety factor of 2.0 for seam IV. However, in seam III, panel IIIA the long-hole deviation resulted in 6.7% variation in the safety factor and therefore the web thickness was suitably adjusted for this panel in seam III. The web thickness given in Table 4.6 was the adjusted one.

As given in Table 4.6, the recovery of coal from seams III and IV increased to about 26% as compared to the earlier design where it was about 20.5%. However,

Table 4.8 Alternate web pillar design for safety factor 2.0 in seam IV and safety factor 1.5 in seam III.

	Panel IV-A/Panel III-A					Panel IV-B/Panel III-B				
	Long holes		Short holes			Long holes		Short holes		
Seam (ht. ext., h)	Avg. length L_L	Web width W_L	Max. length L_S	Web width W_S	% ext	Avg. length L_L	Web width W_L	Max. length L_S	Web width W_S	% ext
IV (4 m)	345 m	13.1 m	115 m	4.8 m	25.7	225 m	14.5 m	135 m	5.5 m	26.8
III (8 m)	340 m	13.1 m	60 m	4.8 m	24.8	220 m	14.5 m	67 m	5.5 m	25.4

Table 4.9 Variations in safety factors with respect to the expected hole deviation for seams IV and III (alternate configuration).

	Panel IV-A/Panel III-A		Panel IV-B/Panel III-B	
Seam	Long holes	Short holes	Long holes	Short holes
IV	±7.7%	±2.4%	±4.7%	±2.7%
III	±6.7%	±1.5%	±4.2%	±1.6%

in case of a web pillar collapse in seam III, it was necessary to assess its impact on the overlying seam IIIA at a parting of about 26 m and seam II at a parting of about 60 m with seam III. The seam IIIA extractable reserve was limited, because panel A was not extractable due to operational difficulties in cutting a thin seam along the apparent dip angle; the reserves were therefore restricted to panel B in seam IIIA.

4.2.8.2 Subsidence impact on seam IIIA

In the event of a single web pillar collapse, the width of the collapsed zone would be 12.5 m in the short holes region, which was non-effective for a parting of 26 m, considering a non-effective width to depth ratio (NEW) of 0.5.

In the long holes region lying at the greatest depth, the width of the collapse zone in the event of a single web pillar collapse would be 21.5 m, which would be greater than a non-effective width for seam IIIA and could cause a sub-surface subsidence. The expected subsidence and strain values were calculated as follows:

Effective mining height $= 1.6$ m (20% ext. of 8 m ht.)
Width of extraction $= 21.5$ m
Depth (parting) $= 26$ m
Predicted maximum subsidence $= 320$ mm
Predicted maximum tensile strain $= 12.3$ mm/m

From recent experiences at the RK-7 mine in the Srirampur area, SCCL, where seam II was being depillared below seam IIA development workings, a maximum tensile strain of 11.5 mm/m was predicted. However, the ground condition did not show any significant deterioration, indicating safe workable conditions. Taking cues from this recent experience, IIIA seam was considered workable even if there was a web pillar collapse in seam III.

When the pillars were designed for a 1.5 safety factor, it was highly unlikely for a cascading pillar failure to occur. In the event of a partial or full pillar failure, the additional load transferred to its neighbouring pillars would be less than 50% of the load originally acting on it. When the neighbouring pillars were designed to take 50% additional load (SF = 1.5), there was little chance of triggering a cascading pillar failure. Earlier cases of cascading pillar failures indicated safety factors of pillars between 1.0 and 1.12.

4.2.8.3 Subsidence impact on II seam

In the event of a web pillar collapse in seam III, the width of the collapsed zone was expected to be about 21.5 m, which would be non-effective for seam II. Therefore, the strategy discussed above for improving the recovery from seams III and IV by keeping a 2.0 minimum safety factor for seam IV and a 1.5 minimum safety factor for seam III was feasible.

Another alternate configuration attempted to improve the extraction in seam III and seam IV would be to go for extraction in seam III by two lifts, leaving a minimum coal parting of 3 m in between. Due to the uncertainty of the local geology and variation in seam thickness, as discussed earlier, there was always a risk of the miner falling into the bottom section while extracting the top section. Therefore, this scheme was only practical if a staggered configuration of extraction cuts were done in the bottom and top seams. Empirical design for such a configuration suggested that it would give a recovery of about 30% from both the seams. However, numerical modelling studies indicated that the staggered configuration would result in much lower overall pillar strength as compared to the columnised pillar configuration, and therefore this option was ruled out. The details of the numerical modelling studies are described serially.

4.2.8.4 Extraction of seam IIIA

Seam IIIA was thinner as compared to the other seams (1.8 m maximum thickness). As discussed above, panel A, in the apparent dip direction was not extractable. This left the option of adopting a suitable web pillar thickness for panel B considering a minimum safety factor of 2.0. The web pillar thickness thus arrived at for seam IIIA panel B is given in Table 4.10. The estimated web pillar thickness was only 5.0 m as given in Table 4.10 and there was therefore practically no scope for having alternate short holes.

The expected variation in safety factor of the web pillars due to hole deviation was within 10% as given in Table 4.10.

4.2.9 Extraction of seam II

Seam II had a maximum thickness of 4.0 m and could be extracted in panel A and panel B. Alternate short and long cuts were designed in this seam for achieving maximum recovery. The design for this seam for a safety factor of 2.0 is as given in

Table 4.10 Web pillar sizes designed for safety factor of 2.0 in seam IIIA.

Full-length holes in panel IIIA-B

Avg. length L_L	Web width W_L	(ht. ext., h)	% ext.	SF variation
190 m	5.0 m	1.8 m	41.2	±9.3%

Table 4.11 Web pillar design for safety factor 2.0 in seam II.

Panel IIA					Panel IIB				
Long holes		Short holes			Long holes		Short holes		
Avg. length L_L	Web width W_L	Max. length L_S	Web width W_S	% ext. (ht. ext., h)	Avg. length L_L	Web width W_L	Max. length L_S	Web width W_S	% ext (ht. ext., h)
275 m	8.1 m	35 m	2.3 m	34.0 (4.0 m)	156 m	8.5 m	40 m	2.5 m	36.6 (4.0 m)

Table 4.12 Variation in safety factors with respect to the expected hole deviation for seam II.

Panel IIA		Panel IIB	
Long holes	Short holes	Long holes	Short holes
±8.5%	±1.4%	±4.8%	±1.6%

Table 4.11. The expected variations in safety factors for seam II extractions are as given in Table 4.12.

4.2.10 Extraction of seam I

Seam I was the top-most seam with no other extractable seams above or surface features to be protected from subsidence. The web pillars in this seam were designed only for short-term stability; say a minimum safety factor of 1.0. All the extractions were to the full length possible. However, this required leaving barrier pillars at regular intervals. The following steps were taken while designing barrier pillars in seam I to prevent cascading web pillar collapse.

1 Barrier pillars would be able to take additional load from the excavated portion on both sides assuming a complete collapse of the web pillars.
2 Wilson's approach (1982) was used to find the abutment loading on to the barrier pillars from the caved goafs. This approach was based on the theory that the full cover pressure in the goaf would occur at a distance from the edge of the barrier pillar equal to 0.3 times the depth of cover.

Table 4.13 Web pillar and barrier pillar design for seam I extractions.

Panel I-A						Panel I-B					
Web pillar	Barrier pillar					Web pillar	Barrier pillar				
Avg. length L_L	Web width W_L	Barrier width W_B	Sub-panel width W_P	No. of web cuts N_C	Overall % ext. (ht. ext., h)	Avg. length L_L	Web width W_L	Barrier width W_B	Sub-panel width W_P	No. of web cuts N_C	Overall % ext. (ht. ext., h)
250 m	3.6 m	15.0 m	75.5 m	11	41.0 (4 m)	138 m	3.9 m	17.0 m	92.3 m	13	39.9 (4 m)

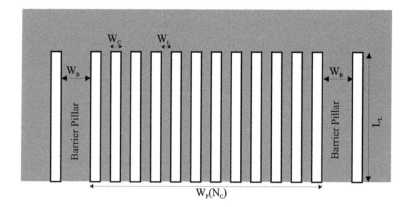

Figure 4.5 Plan showing the scheme for extracting seam I.

3   The barrier pillars would have a minimum safety factor of 2.0, when the web pillars on one side of the barrier were fully crushed forming a caved zone; and a minimum safety factor of 1.5 when both the sides were caved.
4   The spacing between the barrier pillar would be such that the width of each sub-panel would not exceed the critical width resulting in the occurrence of full cover pressure in the middle of the sub-panel, say $W_p < 0.6 \times H$, where W_p is the extraction panel width and H is the depth of cover.

With the above points, the following design was arrived at, as given in Table 4.13.

A schematic plan of the extraction pattern in seam I is given in Figure 4.5. The empirical approach used for the seam I barrier pillar design was a conservative one. Elaborate numerical modelling studies might enable a thinner barrier pillar and improvement in the recovery.

The web pillars were designed for short-term stability and therefore, the web thickness would account for the variation in safety factor due to hole deviation. The expected variation in safety factor was as given below:

Panel IA: ±6.3%

Panel IB: ±11.3%

Table 4.14 Input parameters used in the models.

Rock type	Peak shear strength (τ_{sm}), MPa	Peak friction angle (ϕ_{0m}), °	Tensile strength (σ_t), MPa	Young's Modulus (E), GPa	Poisson's Ratio (ν)	Density kg/m^3	Dilation angle, °
Coal	1.6	29	0.63	3.33	0.25	1500	0
Sandstone	3.2	34	1.3	10.0	0.25	2500	0

The design parameters given in Table 4.13 were the redesigned web pillar thickness as per the above safety factor variation.

4.2.11 Numerical estimation of pillar strength

The web and barrier pillar design in the current study was primarily based on the empirical design approach as discussed above, owing to the greater confidence in the pillar strength equation developed over a couple of decades of observation in a gamut of Indian mining scenarios. However, the empirical design did not really account for aspects such as the behaviour of columnised and staggered pillar configurations in close proximity.

The scope of the numerical pillar strength modelling study was limited to a comparative analysis of the above said pillar configurations in seams III and IV with those of independent workings in seams III and IV. The peak strength values obtained from the numerical modelling exercise were therefore not directly used in the design but used for comparison with independent pillar strength values. Nevertheless, the results would supplement the empirical design by accounting for the variations in pillar configuration for the combined extractions of seams III and IV.

The *FLAC3D* software provides for elasto-plastic analysis of rock excavations with a strain-softening material model using the linear Mohr-Coulomb failure criterion. The strength and elastic constants necessary for numerical modelling using *FLAC3D* in the strain-softening mode are:

1 elastic constants,
2 peak and residual shear strength and its variation with the shear strain,
3 peak and residual angle of internal friction and its variation with the shear strain, and
4 angle of dilation.

All these properties were to be specified for the coal seams and the respective roof and floor rocks. The input parameters used for the strain-softening models were as given in Table 4.14.

The residual values for the cohesion and friction angle were taken as:

$$\tau_{sm} \text{ (residual)} = 0 \tag{4.5}$$

$$\phi_{0m} \text{ (residual)} = \phi_{0m} - 5° \tag{4.6}$$

The variation of the cohesion and friction angles with respect to the shear strain was taken as given in Table 4.15.

Table 4.15 Change in τ_{sm} and ϕ_{0m} with shear strain.

Shear strain	Cohesion (τ_{sm}) MPa	Friction angle (ϕ_{0m}) °
0.000	τ_{sm}	ϕ_{0m}
0.005	$\tau_{sm}/5$	$\phi_{0m} - 2.5$
0.010	0	$\phi_{0m} - 5$
0.050	0	$\phi_{0m} - 5$

Estimation of pillar strength had been made in a way analogous to that of the laboratory estimation of uniaxial compressive strength under servo-controlled testing conditions. However, apart from differences of size between real pillars and laboratory specimens so tested, the latter disregarded *in situ* stresses as well as roof and floor behaviours. The method of modelling consisted of the different steps such as (1) grid generation (2) selection of appropriate model (3) incorporation of material properties, *in situ* stresses and boundary conditions (4) development of roadway excavations in the model and (5) application of a constant vertical velocity at the top of the model and continuous monitoring of the average vertical stress and strain in the pillar at each solution 'step'.

To define the ultimate load-bearing capacity or the peak stress and to obtain the post-failure behaviour of the pillar, a strain-softening material model had been chosen. The relevant material properties and *in situ* stresses employed had already been discussed. After making the roadway excavations, the top of the model was fixed in the vertical direction to maintain a constant vertical velocity. The other boundary conditions included zero vertical displacements at the model bottom and zero normal displacements at the four vertical symmetry planes. The top of the model was fixed in the vertical direction and at a constant velocity of 10^{-5} m/s. The average vertical stresses in all pillar zones and the average vertical strain in the pillar were continuously monitored and plotted. The vertical strain in the pillar was calculated as the average roof-to-floor movement (convergence) over the pillar area divided by the nominal pillar height. It could be noted that the peak value of the strain-stress relation curve defined the pillar strength.

The strain-softening models run for the comparative studies included the following:

Model 1: Independent extraction in seam III to a height of 8.0 m (two-pass).

Model 2: Independent extraction in seam IV to a height of 4.0 m.

Model 3: Simultaneous extraction in seam IV to a height of 4.0 m and seam III to a height of 8.0 m, with a stone and coal parting of 7 m, pillar dimensions corresponding to a 2.0 safety factor in both the seams (Tables 4.6 and 4.7).

Model 4: Simultaneous extraction in seam IV to a height of 4.0 m and seam III to a height of 8.0 m, with a stone and coal parting of 7 m, pillar dimensions corresponding to safety factors of 2.0 and 1.5 in seams IV and III, respectively (Table 4.8 and 4.9).

Model 5: Simultaneous extraction in seam IV to a height of 4.0 m and seam III in two sections to a height of 4.0 m each, with a 5.0 m stone parting between seams IV and III (bottom) and a 3.0 m coal parting between seam III (bottom) and seam III (top), pillar dimensions corresponding to 2.0 safety factors in all the sections.

4.2.12 Results

The loading curves for the above models had been plotted as vertical stress versus vertical strain on the pillars, for the configurations described above. The peak stress obtained for the models is as given below:

Model 1: Seam III extraction, independently: Peak stress = 9.24 MPa
Model 2: Seam IV extraction, independently: Peak stress >16 MPa
Model 3: Seams III and IV extractions, superposed (SF = 2.0 in seams III & IV): Peak stress = 9.51 MPa
Model 4: Seams III and IV extractions, superposed (SF = 1.5 in seam III & 2.0 in seam IV): Peak stress = 8.73 MPa
Model 5: Seam III extraction in two staggered sections and seam IV right below (SF = 2.0): Peak stress = 7.9 MPa

The vertical stress-strain curves along with the plasticity states obtained after crossing the peak stress values are plotted for Models 1–5 as given in Figures 4.6–4.10, respectively. After analysing these loading curves and the peak strength values, the following inferences were drawn:

1   Seam IV pillars, being "squat" for a long web pillar ($W/h > 4$), have much higher post-failure strength and might be "indestructible" for those dimensions. The plasticity states given in Figure 4.7 (Model 2) show a pillar core remaining unyielded even after excessive loading of over 16 MPa.

2   The weaker section would therefore be the seam III working of 8.0 m height. Model 1 (Figure 4.6) shows the independent strength of the seam III pillar to be about 9.24 MPa, which corresponds to an approximate safety factor of 2.0, thereby corroborating the empirical design.

3   Model 3 represents the combined effect of the seam IV and seam III workings with safety factors of 2.0, which also gives a similar strength of 9.51 MPa, indicating that these two seams may be treated as independent. The plasticity states given in Figure 4.8 also confirm the independence of these two workings. The overall strength is very much dependent on the seam III pillar strength.

4   Reduction in the pillar thickness to fit a 1.5 safety factor in seam III and a 2.0 safety factor in seam IV will result in a marginal reduction in the overall strength (about 10% lower). The plasticity states and the loading curve are given in Figure 4.9. Therefore, this configuration is also feasible from a pillar stability point of view.

5   The mutually staggered configuration of the three-section combined workings of seam III and seam IV yielded an overall strength of 7.9 MPa. This is considerably lower (about 20%) than the Model 3 configuration and therefore not recommended from safety point of view. The plasticity states (Figure 4.10) also indicate the yield zones in the vicinity of the seam III workings getting merged to trigger an early collapse of the staggered pillar system.

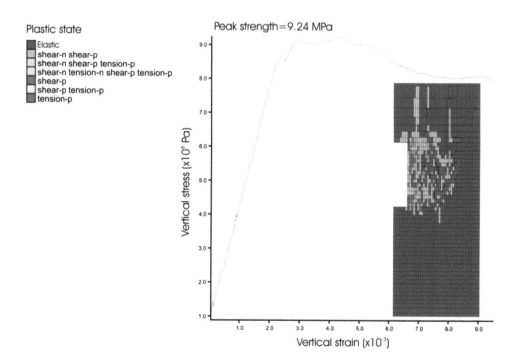

Figure 4.6 Stress-strain curves and plasticity states for Model 1.

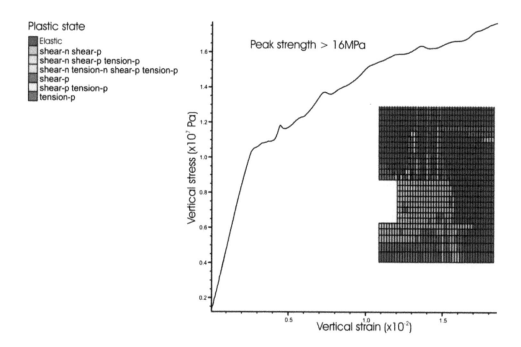

Figure 4.7 Stress-strain curves and plasticity states for Model 2.

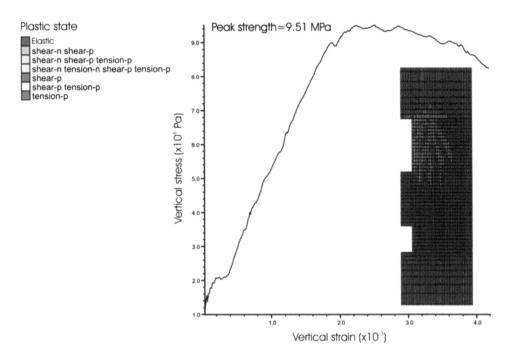

Figure 4.8 Stress-strain curves and plasticity states for Model 3.

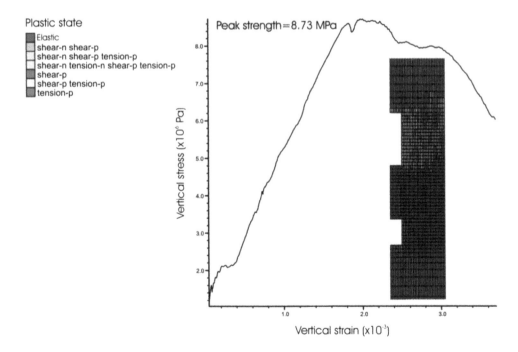

Figure 4.9 Stress-strain curves and plasticity states for Model 4.

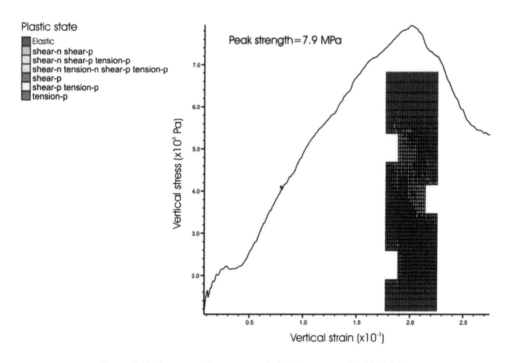

Figure 4.10 Stress-strain curves and plasticity states for Model 5.

4.2.13   Roof stability of web cuts

The roof stability of the web cuts was assessed using numerical modelling. This was essential because during highwall mining, the machine might go up to 350 m under an unsupported span of 3.5 m. For this study, seven representative two-dimensional elastic models were run. The models were constructed taking advantage of symmetry of the cutting pattern for all the seams. Vertical planes of symmetry had been taken, one passing through the middle of the web pillar and the other passing through the middle of the web cut. The geo-mining parameters of the different models are given in Table 4.16.

Numerical modelling was conducted using *FLAC3D*. The safety factor block contours presented were evaluated using a programming environment (*FISH*) of the above software and the theory behind the local safety factor evaluation is presented in Section 4.2.15.

4.2.14   Rock properties used for modelling

Physico-mechanical properties used in the elastic models were taken from NIRM's tested data of core samples of borehole AMT-RGOC2-001 as given in Table 4.17. RMR values for the different types of rock were found to be varying widely; average values taken for coal were 55 and for sandstone was 60.

Table 4.16 Geo-mining parameters of different models for web cut roof stability analysis.

Model no.	Seam(s) & extraction thickness	Positioning of cut	Pillar width (side to side), m	Depth of cover	Immediate roof rock
M-1	IV-4.0 m & III-8.0 m	Seam IV full thickness & seam III along roof	22.5	III-250 m & IV-263 m (long hole)	Sandstone in both workings
M-2	IV-4.0 m & III-8.0 m	Seam IV full thickness & seam III leaving 1.0 m coal against the roof	22.5	III-250 m & IV-263 m (long hole)	1.0 m coal in III & sandstone in IV
M-3	IV-4.0 m & III-8.0 m	IV seam full thickness & seam III along roof	9.5	III-180 m & IV-193 m (short hole)	Sandstone in both workings
M-4	IIIA-1.8 m	full thickness	5	206 m	Sandstone
M-5	II-4.0 m	full thickness	8.5	164 m	Sandstone
M-6	II-4.0 m	full thickness	2.5	50 m	Sandstone
M-7	I-4.0 m	Along floor	4.5	140 m	Carbonaceous sandstone

Table 4.17 Physico-mechanical properties used for the modelling.

Rock type	σ_c MPa	σ_t MPa	Density kg/m³	E GPa	Poisson's ratio	RMR
Sandstone below seam IV	42.25	2.81	2500	6.2	0.25	60
Seam IV coal	40.59	2.71	1500	3.4	0.25	55
Sandstone between seams IV & III	30.9	2.06	2500	10.7	0.25	60
Seam III coal	40.7	2.72	1500	3.2	0.25	55
Sandstone bet. seams III & IIIA	24.6	1.64	2500	10.1	0.25	60
Seam IIIA coal	38.72	2.58	1500	3.5	0.25	55
Sandstone between seams IIIA & II	35.9	2.39	2500	10.1	0.25	60
Seam II coal	38.13	2.54	1500	3.1	0.25	55
Carbonaceous shale between seams II & I	30.6	2.04	2500	2.3	0.25	55
Sandstone between seams II & I	37.45	2.50	2500	10	0.25	60
Seam I coal	34.31	2.28	1500	2.8	0.25	55
Sandstone above seam I	11.07	1.12	2500	5.0	0.25	60

4.2.15   Rock mass failure criterion

The failure criterion for rock mass proposed by Sheorey (1997) had been used for estimating roadway stability, which is given by:

$$\sigma_1 = \sigma_{cm} \left(1 + \frac{\sigma_3}{\sigma_{tm}}\right)^{b_m} \tag{4.7}$$

where σ_1 is the major principal stress required for the failure of rock mass when the minor principal stress is σ_3.

$$\sigma_{cm} = \sigma_c \exp\left(\frac{RMR - 100}{20}\right) \tag{4.8}$$

$$\sigma_{tm} = \sigma_t \exp\left(\frac{RMR - 100}{27}\right) \tag{4.9}$$

$$b_m = b^{RMR/100} \tag{4.10}$$

where
σ_c and σ_t are the laboratory compressive strength and tensile strength, measured in MPa.
RMR is the rock mass rating (Bieniawski's 1976 RMR classification).
b is a constant for the rock type, which is taken as 0.5.
σ_{cm} and σ_{tm} are the rock mass compressive and tensile strength, measured in MPa.
b_m is a constant for rock mass.

4.2.16 Determination of local safety factor

From the output of principal stresses obtained from the numerical models and the above rock mass failure criterion, the safety factor of all elements in the model is defined as:

$$F = \frac{\sigma_1 - \sigma_{3i}}{\sigma_{1i} - \sigma_{3i}} \tag{4.11}$$

except when $\sigma_{3i} > \sigma_{tm}$

$$F = \frac{\sigma_{tm}}{-\sigma_{3i}} \tag{4.12}$$

In the above equations, σ_{1i} and σ_{3i} are the major and minor induced stresses from the numerical model output. The sign convention followed here is negative for tensile stresses and positive for compressive stresses.

4.2.17 Modelling procedure

Numerical modelling studies had been done in the following stages:

Stage 1: The 2D models were constructed using the vertical planes of symmetry (half width of web cut and half width of pillar). The virgin stage was modelled loaded with the elastic properties of the different rock types.

Stage 2: Roadway of half width and full height was excavated and the model was run to attain equilibrium.

Stage 3: The FISH program was run using the output of the numerical model along with strength parameters of different rock types to analyse the web cut roadway stability in terms of safety factor contours.

4.2.18 Modelling results

Results of the representative numerical models are given below.

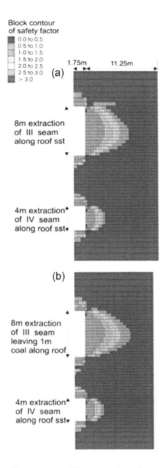

Block contour
of safety factor

0.0 to 0.5
0.5 to 1.0
1.0 to 1.5
1.5 to 2.0
2.0 to 2.5
2.5 to 3.0
> 3.0

(a)

1.75m   11.25m

8m extraction
of III seam
along roof sst

4m extraction
of IV seam
along roof sst

(b)

8m extraction
of III seam
leaving 1m
coal along roof

4m extraction
of IV seam
along roof sst

Figure 4.11 Roof stability of seam III and seam IV extractions for long holes (a) 8.0 m along the roof, (b) 8.0 m leaving 1.0 m coal along the roof.

4.2.18.1 Model-1 and Model-2

These models represented the roof stability of the combined excavations of seams III and IV. In Model-1, the seam III extraction was taken along the roof and in Model 2 it was taken 1.0 m below the sandstone roof, leaving 1.0 m of coal. The depth and pillar dimension represented the long-hole region existing at the highest depth.

The block contours of the safety factors obtained for Model 1 configuration are shown in Figure 4.11(a). From the plot it is clear that at the centre of the roadway of seam III, the safety factors are between 0.5 and 1.0 up to a height of 1 m. On the other hand in seam IV, the immediate roof showed safety factors that are more than 1. Model 2 results as shown in Figure 4.11(b) showed considerable improvement in the roof safety factors, due to the 1.0 m coal left against the weaker roof of seam III.

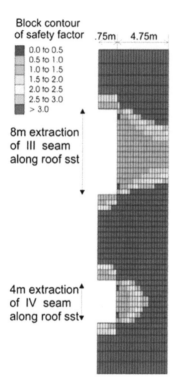

Block contour
of safety factor

- 0.0 to 0.5
- 0.5 to 1.0
- 1.0 to 1.5
- 1.5 to 2.0
- 2.0 to 2.5
- 2.5 to 3.0
- > 3.0

8m extraction
of III seam
along roof sst

4m extraction
of IV seam
along roof sst

Figure 4.12 Roof stability of seam III and seam IV extractions corresponding to short holes.

4.2.18.2 Model-3

This model represented the short-hole region of the seam III and seam IV workings. The maximum depth and web width modelled were those occurring at the end of the short holes. Figure 4.12 shows these workings, with roof safety factors more than 1.0, owing to the lower depth of cover.

4.2.18.3 Model-4

In this model, seam IIIA extraction had been simulated at the highest depth. Block contours of safety factors obtained for this configuration are shown in Figure 4.13. From the plot it is clear that at the centre of the roadway, safety factors are more than 1.0, indicating immediate roof stability.

4.2.18.4 Model-5 and Model-6

Models 5 and 6 simulated the long and short web cuts in seam II, respectively. Block contours of the safety factor over the web cuts against 8.5 m and 2.5 m wide web pillars, corresponding to long and short-hole regions, at a depth of cover of 164 m and 50 m are shown in Figure 4.14(a) and (b), respectively. From both the model results, it can be seen that the immediate roof strata is showing safety factors of more than 1.0.

1.75m 2.5m

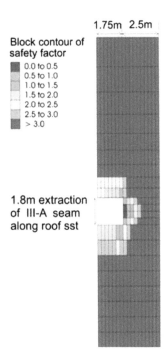

Block contour of
safety factor

0.0 to 0.5
0.5 to 1.0
1.0 to 1.5
1.5 to 2.0
2.0 to 2.5
2.5 to 3.0
> 3.0

1.8m extraction
of III-A seam
along roof sst

Figure 4.13 Roof stability of seam IIIA extractions.

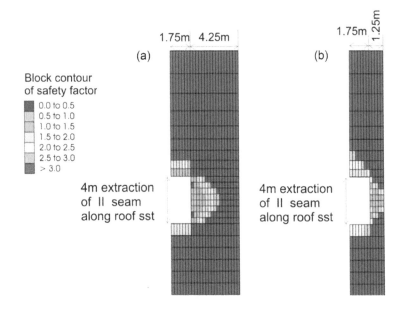

1.75m 4.25m

(a)

1.75m 1.25m

(b)

Block contour
of safety factor

0.0 to 0.5
0.5 to 1.0
1.0 to 1.5
1.5 to 2.0
2.0 to 2.5
2.5 to 3.0
> 3.0

4m extraction
of II seam
along roof sst

4m extraction
of II seam
along roof sst

Figure 4.14 Roof stability of seam II extractions corresponding to (a) long holes, (b) short holes.

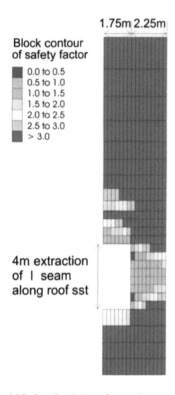

Figure 4.15 Roof stability of seam I extractions.

4.2.18.5 Model-7

Seam I extractions were simulated in Model 7 for 4.5 m wide web pillars at the maximum depth of cover. Block contours of the safety factor are shown in Figure 4.15. Though the immediate roof strata of seam I was weaker compared to the lower seams, it shows safety factors more than 1.0, due to the lower depth of cover.

From the above elastic modelling studies it is clear that the 3.5 m wide cuts would have the required stability during the extractions in all the seams. It was recommended to leave 1.0 m coal against the seam III roof for better stability.

The above stability analysis was purely static. The dynamic effect of opencast mine blasting in the nearby benches on the roof stability of these extraction cuts is discussed in Section 4.2.22.

4.2.19 Entry mouth stability

From past experiences in the USA and Australia, several cases of web pillar collapses at the entry mouth had been seen. The primary reasons could be very steep highwall angles and insufficient web thickness at the entry mouth. When the highwall slope is very steep (70–90°), which is the case in most of these mines, it tends to throw a higher cover pressure right at the beginning of the entry. The web pillars have three free faces at the entry mouth, which makes the webs vulnerable to spalling and crushing at the mouth.

At OCP-II, the highwall slope was much gentler at about 47°. At the entry mouth region the cover pressure was very nominal and therefore the entries were expected to be very stable.

4.2.20 Recommendations

From the empirical and numerical modelling design studies explained in the preceding sections, the following recommendations were made for safe extraction of all the five targeted seams at OCP-II by highwall mining:

1. The highwall mining block was generally divided into panel A and panel B, the extraction cuts were planned in these panels along apparent dip and full dip directions, respectively.
2. Except in seam IIIA, all the seams were to be extracted in panel A and panel B. In seam IIIA, only panel B was proposed to be extracted.
3. To find whether web pillars had collapsed in the extracted seams, it was advised to monitor the ground movements from the surface.
4. Extractions in each seam could take several months and therefore spontaneous heating of the exposed coal web pillars and the occurrence of fire may take place; this needs to be monitored.

The extraction design parameters recommended for individual seams are as given below.

4.2.20.1 Seams III & IV combined

1. Seam IV was to be extracted to a maximum height of 4.0 m from the floor.
2. Seam III was to be extracted to a maximum height of 8.0 m, in two passes, in descending order, after leaving 1.0 m thick coal against the stone roof of seam III for better roof stability.
3. The seam IV and seam III workings were to be kept superposed.
4. Alternate short and long extraction cuts were proposed in both the seams, superposed to each other. However, the length of the short holes varied in both the seams to achieve maximum recovery.
5. Two options suggested for the web pillar design in these seams were.

   Option 1: with a safety factor of 2.0 in both the seams, 20.5% extraction (Table 4.6), and
   Option 2: with a safety factor of 2.0 in seam IV and a safety factor of 1.5 in seam III, 26% extraction (Table 4.8).

   The mine management could choose between these.
6. The pillar sizes and lengths of the holes were as per Table 4.6 and Table 4.8 for Option 1 and Option 2, respectively.

4.2.20.2 Seam IIIA

1. Extraction was proposed only in panel IIIB to a maximum height of 1.8 m.
2. All extraction holes were proposed as full length.

3   The web pillar size was designed for a minimum safety factor of 2.0 as given in Table 4.10.
4   A 41.2% extraction was projected from panel IIIB.

4.2.20.3 Seam II

1   A maximum of 4.0 m height of extraction was proposed.
2   Alternate short and long extraction cuts were proposed for maximising the recovery.
3   The web pillar thickness required for a minimum safety factor of 2.0 is given in Table 4.11.
4   An average extraction of about 35.5% was expected from panels IIA and panel IIB.

4.2.20.4 Seam I

1   A maximum height of 4.0 m could be extracted in seam I, preferably along the floor of the seam for a better roof condition.
2   All extraction cuts were designed to full length.
3   The web pillars were designed for a minimum safety factor of 1.0, considering short-term stability due to the absence of further workable seams and important surface features. The web pillar design also considered the maximum hole deviation.
4   Barrier pillars were also designed after a specific number of extraction cuts.
5   The widths of the web and barrier pillars in panel IA and panel IB along with the barrier pillar intervals are given in Table 4.13.
6   An average extraction of about 40.5%, considering the barrier pillars, was projected from panel IA and panel IB.

4.2.21 Suggestions for future highwall mining operations

It may be said that every opencast coal mine, theoretically, has a scope for highwall mining towards the end of its life, so why not plan in advance? In most of the Indian coalfields, multiple coal seams to exist. Once the open pit reaches its limit and the final slope is formed, it will not be possible to make benches for working the upper seams. This will leave us with the option of extracting the seams by highwalling in ascending order, after leaving a considerable amount of coal as support pillars for the top seams. Had these seams been worked in a descending order the recovery could be significantly improved. As in the case of OCP-II, the extraction percentage is quite lower for the bottom seams due to the ascending sequence. To minimise such losses, a few suggestions are made:

1   When the open pit nears its boundaries and being worked with benches, highwall operations may be planned in descending order. The working benches will provide a platform for the extraction of the remaining portion of the seams.
2   The highwall operations may go parallelly with the open pit final slope formation in descending order.
3   A suitable backfilling technology may be developed in order to improve the recovery and ground condition.

Table 4.18 Threshold values of vibration (measured on the roof) for the safety of roofs in the underground workings for different RMRs.

RMR of roof rock	Threshold values of vibration in terms of peak particle velocity (mm/s)
20–30	50
30–40	50–70
40–50	70–100
50–60	100–120
60–80	120

[DGMS (Tech.), 2007].

Table 4.19 Threshold values of vibration (measured on pillars) for the safety of roofs in the underground workings for different RMRs.

RMR of roof rock	Threshold values of vibration in terms of peak particle velocity (mm/s)
20–30	20
30–40	20–30
40–50	30–40
50–60	40–50
60–80	50

[DGMS (Tech.), 2007].

4   A suitable highwall cutter may be used such that it takes a narrow cut while entering and widens the cut while retreating by thinning down the web pillar on both sides. This can give better ground conditions and can also improve recovery.

4.2.22 Impacts of surrounding blasting on highwall mining

Blasting activities close to the proposed highwall mining at OCP-II could affect the stability of highwall mining entries, web pillars as well as the highwall of the opencast where from the workings are to be executed. Ground vibration generated from the day-to-day blasting operation nearby the highwall mining entries could impose premature roof and side collapse, thereby trapping the continuous miner inside the highwall mining entries. Flyrock generated from blasting nearby the highwall mining could also damage the equipment used for the mining like the launch vehicle etc. and cause safety concern to the workers.

The Directorate General of Mines Safety of India (DGMS) stipulates the threshold values of vibration for the safety of roofs and pillars in underground coal mines. The safe values of ground vibrations in the roof of galleries (at the junction points) in terms of peak particle velocity (PPV) based on Rock Mass Rating (RMR) values are given in Table 4.18. The limiting values of PPV in the pillars are given in Table 4.19 (DGMS Technical Circular No. 06 of 2007).

Table 4.20 Damage level of rock mass based on ground vibration (after Bauer & Calder, 1971).

Particle velocity (mm/s)	Predictable damages
<250	No danger in sound rock
250–600	Possible sliding due to tensile breakage
600–2500	Strong tensile and some radial cracking
>2500	Complete break-up of rock masses

Blast-induced ground vibrations may also affect the stability of the highwall slope. Ground vibrations have twofold actions on rock mass. On the one hand, they affect the integrity of the rocks or their strength parameters and, on the other, can provoke wall or slope collapses when destabilising actions are introduced (Jimeno et al., 1995). During the process of blasting, the energy that is not used in the fragmentation and displacement of rock, which is sometimes more than 85% of that developed in the blast, reduces the structural strength of the rock mass outside the theoretical radius of the action of excavation. New fractures and planes of weakness are created and joint, declasses and bedding planes that initially behaved non-critically, when opened, result in an overall reduction of rock mass cohesion (Pal Roy, 2005). This is manifested in overbreak, leaving fractured mass in a potential state of collapse. Bauer and Calder (1971) proposed the following generalised criteria, as given in Table 4.20, for the damage level of particle velocity on rock mass and slopes.

4.2.22.1   Field study and ground vibration results

The Blasting Department (presently Rock Excavation Engineering Division) of CSIR-CIMFR carried out field investigations in RG OCP-II to assess the impact of ground vibration on the stability of the punch entries of the Adriyala Shaft Project (CSIR-CIMFR, 2008a). The site was the same as that of highwall mining. The research team monitored ground vibrations generated due to twelve production blasts at different working benches of the OCP-II. The total explosive charge fired in a round of a blast varied between 1.138 and 17.5 tonnes. The maximum charge per delay varied between 91.00 and 673.80 kg. The blasthole diameters used were 150 mm and 250 mm. Hole depth varied between 5.0 and 14.0 m. The nonel system was used for in-hole as well as surface initiation in all the OB blasts at RG OCP-II. The distances of vibration monitoring points for the different blasts varied between 75 and 670 m.

47 ground vibration data were collected during the period of the field investigation. The magnitude of ground vibrations recorded varied between 1.50 and 154.0 mm/s. The maximum magnitude of vibration recorded during the period of field investigation was 154.00 mm/s at a distance of 75 m from the blast site with associated dominant peak frequency of 30 Hz. The maximum explosives weight per delay was 673.80 kg and the total explosives fired in the round was 17,519.00 kg. The frequency levels recorded from the different blasts varied widely, ranging from 6.0 to 39 Hz. Fast Fourier Transform (FFT) analyses of vibration data revealed that in the majority of the blasts, the concentration of vibration energy ranged between 8–24 Hz.

Table 4.21 Nature of roof and corresponding RMR value (High & Low) of different seams at OCP-II, SCCL.

Name of Seam	Seam thickness [m]	Nature of the immediate roof	RMR values	
			Low	High
I	5.2–6.7	Coal & sandstone	33	74
II	3.1–4.0	Sandstone	40	74
IIIA	0.9–1.8	Sandstone	38	79
III	11.0–12.2	Coal & sandstone	33	70
IV	3.7–4.1	Sandstone	33	65

4.2.22.2 Assessment of safe value of PPV

As per DGMS, 2007 (Tables 4.18 and 4.19), the permissible limit of peak particle velocity for the stability of the roof and web pillars of the proposed highwall mining depends on the RMR values of the immediate roof. The nature of the roofs and their respective RMR values for different seams of highwall mining are given in Table 4.21.

The lowest RMR value for the different coal seams was 33. Seam nos. I, III and IV had lower values of RMR (i.e. 33 each). Therefore, taking the lowest RMR value, the threshold values of peak particle velocity for the safety of highwall mining entries came in the range of 50–70 mm/s. Taking the lower value of PPV as the threshold value for better safety, 50 mm/s was considered as the safe value of PPV for different roofs of the highwall mining entries. Based on the generalised criteria proposed by Bauer and Calder (1971) for damage level of particle velocity on rock mass and slopes, the threshold values of PPV (i.e. 50–70 mm/s) were considered to be safe for highwall slopes. The highwall slopes would also have less damage impact due to ground vibration in comparison to the roofs of the galleries.

4.2.22.3 Assessment of safe blasting parameters

Values higher than the threshold PPV were recorded while using 250 mm blasthole diameter and longer hole depths. Flyrock was also observed during the field investigation in most of the blasts. The ground vibration data recorded during the field investigations were grouped together for statistical analysis. An empirical equation was established correlating the maximum explosive weight per delay (Q_{max} in kg), distance of vibration measuring transducers from the blast face (D in m) and recorded peak particle velocity (V in mm/s). The regression plot of vibration data recorded at various locations due to blasting at RG OCP-II, SCCL is given in Figure 4.16. The established equation for the site was:

$$V = 1114[D/\sqrt{Q_{max}}]^{-1.44} \tag{4.13}$$

Coefficient of determination $= 0.882$
Standard deviation $= 0.201$

The above equation was site-specific and is applicable for the RG OCP-II only. This equation could be used to compute the safe maximum explosive weight

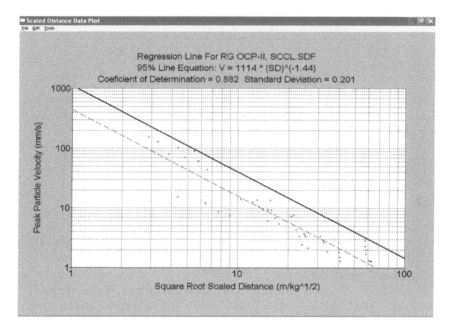

Figure 4.16 Regression plot of vibration data recorded at various locations due to blasting at RG OCP-II, SCCL.

Table 4.22 Safe explosive charge per delay at various distances of concern from the blasting site, keeping the threshold value of ground vibration as 50 mm/s.

Distance of blasting face from the concerned structure [m]	Suggested maximum charge per delay [kg]
50	34.00
100	135.00
150	304.00
200	540.00
250	844.00
300	1216.00
350	1655.00
400	2161.00
450	2735.00
500	3377.00

to be detonated in a delay for distances of concern. Keeping in view of the peak particle velocity of 50 mm/s as the safe level of ground vibration for the total safety of highwall mining entries, the safe levels of maximum explosive weight per delay had been calculated for various distances of concern, using Equation (4.13) and are given in Table 4.22. The same is represented in Figure 4.17.

4.2.22.4 Conclusions

The following conclusions were drawn on the basis of the field study at RG OCP-II for the stability of highwall mining entries, web pillars and highwall slopes from blast-induced ground vibrations:

1   Blasting activities close to the proposed highwall mining at RG OCP-II might affect the stability of highwall mining entries, web pillars as well as the highwall of the opencast wherefrom the workings were to be executed.
2   12 production blasts were monitored during the field investigation at RG OCP-II with varying design and charging patterns. The total explosive charge fired in a round of blast varied between 1.138 and 17.5 tonnes. The maximum charge per delay varied between 91.00 and 673.80 kg. The blasthole diameters used were 150 mm and 250 mm. The hole depth varied between 5.0 and 14.0 m.
3   The lowest value of RMR from the different coal seams in RG OCP-II was 33. With this value of RMR, the threshold values of peak particle velocity for the stability of the roof of the highwall mining entries came in the range of 50–70 mm/s as per DGMS, 2007. Taking the lower value of PPV as the threshold value for better safety, 50 mm/s was considered as the safe value of PPV for roofs of the highwall mining entries.
4   Based on the generalised criteria proposed by Bauer and Calder (1971) for damage level of particle velocity on rock mass and slopes, the threshold values of PPV (i.e. 50–70 mm/s) were considered to be safe for highwall slopes. The highwall slopes had a lesser possibility of damage impact due to ground vibration in comparison to the roofs of the galleries.
5   The propagation equation for prediction of blast-induced ground vibration at RG OCP-II had been established (Equation (4.13)). The recommended explosive weights per delay for various distances from the highwall mining are given in Table 4.22 and Figure 4.17.
6   More experiments are needed for better predictability of blast vibrations.

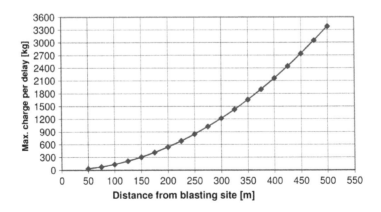

Figure 4.17 Recommended charge per delay to be fired in a round of blast for the safety of highwall mining entries at RG OCP-II.

4.2.23 Slope stability analysis

This section deals with the slope stability study of the highwall at the OCP-II mine. The stability of slopes depends on the geological structures, geo-mechanical properties of the slope materials and groundwater/rainwater conditions in the quarry. The slope stability analysis is carried out by the limit equilibrium method after identifying kinematically possible failure modes. Circular failure is the basic mode of slope instability in weathered slope material, whereas plane – wedge – toppling failures occur in the hard rock mass (Figure 4.20).

The rock discontinuities were mapped at the exposed benches of the pit as per the norms of the International Society of Rock Mechanics (ISRM, 1978). Geo-technical mapping was undertaken to determine the critical orientation of structural discontinuities. With the available geo-technical information related to the highwall slope, the design was made for a stable slope at OCP-II.

4.2.23.1 Geo-hydrology

Well fractured rock mass coupled with existing opencast and underground workings in close proximity had made the geo-mining conditions of the quarry into a drained condition for all practical purposes after implementing an effective drainage system. If the slope mass was not provided with an effective drainage system, i.e. in undrained-conditions, a phreatic surface would most likely develop in the sandstone below the top soil.

4.2.23.2 Geo-mechanical properties

It is prudent to know the lithological units in which the slope is to be cut. The engineering properties of these lithological units will always influence the analysis of the slope stability. The geo-mechanical properties of various rock layers present in the mine slope are presented in Table 4.23.

Table 4.23 Geo-mechanical properties of different lithology of active mine slope material.

Sr. No.	Lithology	Cohesion (kPa)	Friction angle (°)	Density (kN/m³)
1	Loose soil	26.00	26.0	16.50
2	Black cotton soil	30.00	20.0	17.00
3	Parting (Grey SandStone GSS)	310.00	33.0	25.30
4	Coal seam I	172.00	27.0	13.40
5	Sandstone parting (GSS)	305.00	33.0	24.20
6	Coal seam II	172.00	27.0	13.40
7	Sandstone parting (GSS)	305.00	32.0	23.40
8	Coal seam IIIB	160.00	27.0	13.20
9	Sandstone parting (GSS)	303.00	32.0	23.20
10	Coal seam IIIA	160.00	27.0	13.20
11	Sandstone parting (GSS)	305.00	33.0	25.00
12	Coal seams III & IV	180.00	28.0	13.30
13	Sandstone parting (GSS)	305.00	33.0	25.00

Table 4.24 Orientation of major joint sets in the sandstone.

Joint sets	Mean orientation of joint sets		Spacing (cm)	Persistence (m)
	Dip direction (degree)	Dip amount (degree)		
J1	N130	86	150	2.5
J2	N050	14	100	5.0
J3	N232	84	150	1.0

J1 is a systematic joint set. J2 and J3 are sub-systematic joint sets (ISRM, 1978).

Table 4.25 Orientation of cleats in the coal.

Joint sets	Mean orientation of cleats		Spacing (cm)	Persistence (m)
	Dip direction (degree)	Dip amount (degree)		
CL1	N050	14	20	0.6
CL2	N155	86	10	1.0
CL3	N260	84	40	0.6

4.2.23.3 Physical characterisation of discontinuities

Geo-technical mapping was done on the exposed benches of OCP-II. The mapping was done as per the norms of the ISRM (1978). It may be mentioned here that the characteristics of discontinuities were varying in different parts of the pit.

The general structural pattern in the rock mass is presented in Table 4.24.

Many discontinuities, observed in the field, were open due to the effect of poor blasting. Other random and less persistent fractures were also present. But these were of less importance from a stability point of view due to their lower persistence. The dip amount and dip direction of fractures in coal seams (cleats) are shown in Table 4.25. Though it shows modal values, slight deviation had also been observed.

4.2.24 Slope stability analysis

The stability analysis was done using the *GALENA* computer programme, which is based on the limit equilibrium method. The cut-off safety factor for stability was considered to be 1.3.

4.2.24.1 Kinematic analysis

The average orientations of the discontinuity sets determined from the geological structural mapping were analysed to assess kinematically possible failure modes involving structural discontinuities (Figures 4.21 and 4.22).

The kinematic analyses, to determine the types of failure in the sandstone as well as coal slope faces, are presented in Figures 4.21 and 4.22. The figures shows that the

Table 4.26 Stability analyses of highwall slope.

Existing highwall slope parameters	Factor of safety
Maximum overall slope height of highwall = 220 m Maximum overall slope angle of highwall = 47°	Drained-slope – 1.3 (Figure 4.23)

J3 joint set is striking parallel to the NW-SE slope face but dipping at an angle greater than the overall slope face. So, large-scale failure along these joint sets (Tables 4.22 and 4.23) were unlikely.

As far as wedge failure was concerned, the wedge geometry formed by the intersection of the J1 and J3 joint sets, was striking approximately parallel to the slope face but the wedge was dipping at an angle more than the slope face. So, this wedge geometry would not get exposed on the slope face (Figure 4.21), hence large-scale wedge failure was unlikely.

The kinematic analyses showed that large-scale failure was unlikely but small-scale wedge instability could not be completely ruled out, which might be formed by the randomly-oriented joints present in the slope face especially during the monsoon season.

4.2.24.2 Stability analysis by the limit equilibrium method

The stability analysis was conducted along a representative cross section, which showed the maximum overall slope height and maximum overall slope angle of the standing highwall. The safety factor obtained for the 220 m high and 47° highwall slope was 1.3 (Table 4.26). As mentioned previously, the most likely geo-mining condition at the OCP-II would be a drained-slope condition. The analysis was done along a typical cross section provided by the mine management (Figure 4.23).

The stability analysis shows that large-scale slope instability was unlikely. But the highwall slope was showing critical wedges and weak planes in the standing highwall slope. It was taken care of by applying suitable remedial measures.

Implementing the extraction design of CSIR-CIMFR, the first highwall mining operation in India started on 10 December 2010 at OCP-II, RG-III area, SCCL, and it extracted around 0.15 million tonnes of coal from seam I and seam II (Figures 4.18 and 4.19). Extraction in seam II had problems due to unexpected geological discontinuities and poor roof conditions. However, in seam I, the extractions went smoothly with almost all the holes being extracted to full strength as per the design provided by CSIR-CIMFR. Almost 1/5th of the total project cost of highwall mining was recovered before the machine moved to the second nearby mine namely Medapalli Opencast Project of SCCL during the end of September, 2011 where it operated safely for the extraction of six seams based on the design provided by CSIR-CIMFR.

4.2.24.3 Remedial measures

The proper levelling of the top ground surface of the opencast project is necessary to divert the maximum rainwater of the upper surface away from the pit. Otherwise, soil

Figure 4.18 Inauguration of highwall mining system at OCP-II, RG-III area, SCCL on 10 December 2010.

Figure 4.19 Production of coal from highwall mining at OCP-II, RG-III area of SCCL.

erosion and gully erosion may cause small-scale collapses in the working area, which will endanger the miners and machinery working at the lower level in such areas.

The excavated pit is to be provided with an effective garland drain. The drainage must always be directed away from the excavated pit. It is necessary to avoid the flow of rainwater from the adjacent catchment area to the benches of the precarious lithological units, which affect their slope stability. The drain should be kept clear of soil debris and effective for the free flow of water.

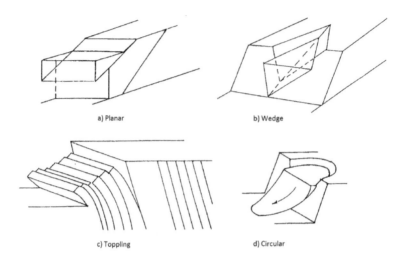

Figure 4.20 Modes of common slope failures (a) planar (b) wedge (c) toppling (d) circular (Hoek and Bray, 1981).

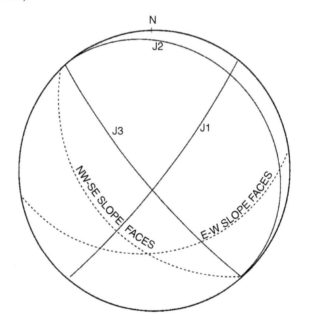

Figure 4.21 Kinematic analysis of types of failure in sandstone slope faces.

If damage due to poor blasting is visible on the standing highwall slope faces, loose dressing of the overbreaks, overhangs etc. should be done.

At least two slope monitoring observations should initially be taken at an interval of one month. The two consecutive observations would reveal whether there is movement or not. The mine management could do it by installing monitoring stations at

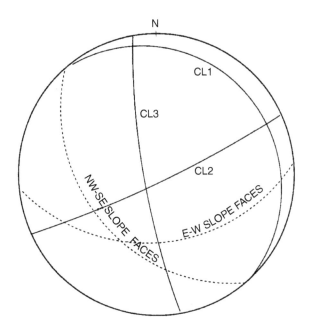

Figure 4.22 Kinematic analysis of types of failure in coal slope faces.

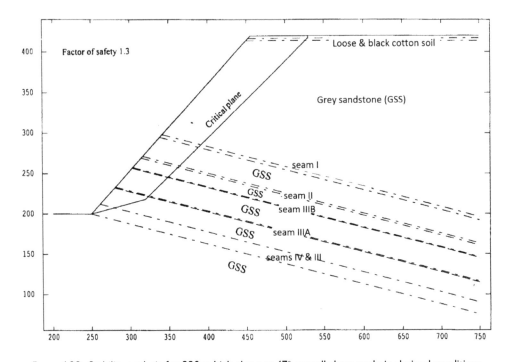

Figure 4.23 Stability analysis for 220 m high slope at 47° overall slope angle in drained conditions.

an interval of about 25 m above the zone of interest of the highwall slope. The total station instrument available with the survey department could be used for this purpose.

4.2.24.4 *Findings and suggestions*

1   An assessment of the engineering and structural geology, strength properties and the related geo-technical controls indicated that the large-scale slope instability in the highwall was unlikely but small-scale wedge instability could not be completely ruled out. The slope monitoring was done to detect any movement in any part of the standing highwall slope.

2   Supported/reinforced wire-mesh, after a thorough loose dressing, might be installed near the zone which was observed to be vulnerable to rock fall because practically wire-netting the entire highwall was not feasible. It would minimise the rock fall. Moreover a protective canopy between the launch vehicle and the high-wall would further improve safety. A protective bund was suggested at the working level to stop the small boulders falling from the lower portion of the standing highwall slope.

3   A bund or garland drain or a combination of the two was proposed depending upon the geo-morphology/topography of the area, at the highwall crest to avoid the inrush of surface rain water into the pit. The existing tension cracks were to be filled and sealed properly before the onset of the monsoon. The drain should be properly graded to promote quick water movement and minimise the chances of ponding. The drain should be kept clean.

4.3 SITE B

HIGHWALL MINING AT MEDAPALLI OPENCAST PROJECT OF M/S SINGARENI COLLIERIES COMPANY LIMITED (SCCL)

At Medapalli Opencast Project (MOCP) of Singareni Collieries Company Ltd. (SCCL), highwall mining was proposed to extract the left-out coal beneath the final pit slope (highwall). SCCL had contracted with Advanced Mining Technology Pvt. Ltd. (AMT) and ADDCAR Highwall Mining Systems of the USA to extract the exposed coal seams at the highwall: seam I, seam II, seam III, seam IIIA, seam IV and seam V, occurring in descending order. The coal seams were easterly dipping with a maximum dip of about 1 in 6 (9°).

The mine lease boundary abutted at the Godavari river bank and the open-pit had reached its final pit limit at about 140–150 m away from the river bank. A flood protection bund separated the pit from the river.

The coal existing between the final pit wall and the river bank, within the mine lease, was available for highwall mining, since it could not be extracted by any conventional mining methods. A safe distance corresponding to the angle of draw in all the seams was to be left against the river bank.

Towards the northern and eastern sides of the pit, the seam exposed at the highwall was proposed for highwall mining.

Figure 4.24 Medapalli OC Project and the final pit slope.

Figure 4.25 Photograph showing a closer view of the highwall.

Approximately a 350 m to 1100 m length of highwall was available for extraction in all the seams. A major fault separated the proposed region into two blocks. At the up-throw side of the fault, towards the north of the pit, seams V, IV, III and IIIA were targeted in block A. Towards the down-throw side, towards the east side of the pit, seams V, IV, III, IIIA, II and I were available for highwall mining, demarcated as block B.

A photograph showing the highwall at Medapalli OCP is given in Figure 4.24. The wall towards the right, which was roughly in the north-eastern direction of the pit, was targeted for highwall mining. A closer view of the highwall is given in Figure 4.25.

4.3.1 Geo-mining details

The geo-mining parameters related to the highwall mining blocks had been supplied by AMT and SCCL. A plan showing the highwall mining blocks, namely block A at the northern side and block B towards the east is given in Figure 4.26. From the figure,

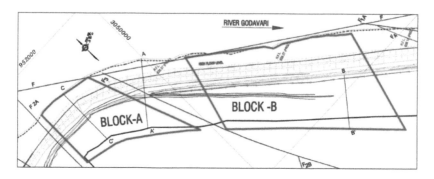

Figure 4.26   Plan showing the highwall mining blocks A and B.

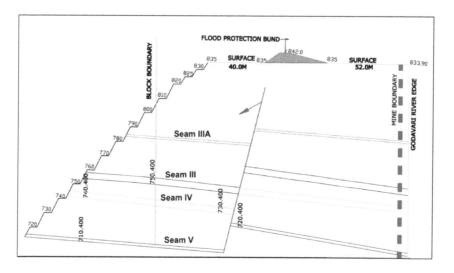

Figure 4.27   Section along A–A' showing the fault separating the blocks.

the fault separating the two blocks can also be seen. In fact, the blocks were demarcated by leaving a sufficient distance from the fault. A vertical section A–A' passing through block A, showing the fault separating the two blocks, is given in Figure 4.27; the section shows the position of the current pit, ultimate pit, flood protection bund and the river bank. The targeted seams in block B are shown along a vertical section B–B' in Figure 4.28.

The seams occurring in block A are shown along a vertical section C–C' in Figure 4.29. Section C–C' also depicts the calculation of the safe distance according to an angle of draw of 25° from the river bank. The lithology of the coal measure rocks at the site and the seams is shown along borehole sections in Figure 4.30.

Details of the seam thickness, depth of cover and intervening vertical parting between the workable seams are presented in Table 4.27 for block A and block B.

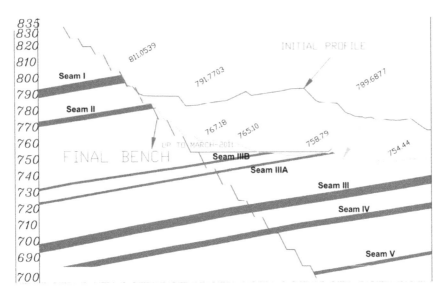

Figure 4.28 Section along B–B' showing the seams in block B.

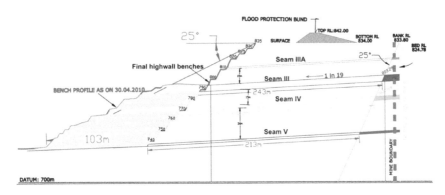

Figure 4.29 Section along C–C' showing the seams in block A, flood protection bund, river bank and the concept of angle of draw for limiting the workings.

As seen from Table 4.27, the intervening parting at all instances is sufficiently thick (>9.0 m), and therefore the contiguous seam scenario did not arise and all seams could be treated separately in the extraction designs.

Physico-mechanical properties of the coal measure rocks were previously tested by the National Institute of Rock Mechanics (NIRM) in 2007 (NIRM, 2007). The tested values included uniaxial compressive strength, Brazilian tensile strength, density, Young's modulus, Poisson's ratio and slake durability. These tested properties for intact rock, for the rock types, including the coal seams and their roof and floor, are given in Table 4.28.

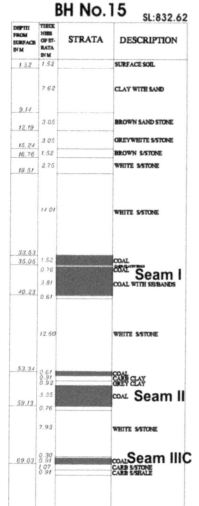
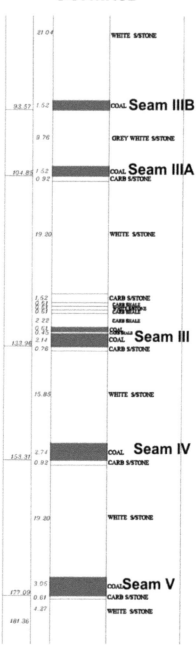

Figure 4.30 A representative borehole section showing the seams in the proposed blocks.

Table 4.27 Seam details.

Seam	Depth of cover (m)	Thickness (m)	Vertical parting (m)	Remarks
I	26–71	2.5–4.5	–	Only in block B
II	25–90	2.5–3.0	16–17 with seam I	Only in block B
IIIA	27–141	1.5–2.0	43–47 with seam II	In blocks A and B
III	30–170	5.0–6.5	21–25 with seam IIIA	In blocks A and B
IV	44–192	3.0–5.0	11–12 with seam III	In blocks A and B
V	83–224	1.0–2.0	20–30 with seam IV	In blocks A and B

Table 4.28 Physico-mechanical properties.

Strata	σ_c (MPa)	E (GPa)	ν	Density (kg/m^3)	σ_t (MPa)	Slake durability index (%)
Seam I roof	32.7	22.6	0.27	2600	0.8	80
Seam I	40.3	2.5	0.26	1500	3.9	92
I-II parting	12.0	4.0	0.19	2600	1.5	91
Seam II	33.0	2.7	0.19	1500	3.7	80
II-IIIA parting	20.0	7.0	0.25	2600	2.4	91
Seam IIIA	39.1	3.1	0.25	1500	3.9	94
IIIA-III parting	45.1	5.7	0.25	2600	2.1	76
Seam III	36.3	3.4	0.32	1500	3.6	–
III-IV parting	40.4	12.0	0.3	2600	4.0	90
Seam IV	36.3	2.7	0.3	1500	2.1	97
IV-V parting	54.7	10.0	0.3	2600	1.7	–
Seam V	38.5	2.8	0.33	1500	3.8	84

Table 4.29 Rock mass properties of roof.

Seam	RQD	RMR	Q
I	73	57	4.2
II	82	64	9.2
IIIA	70	55	3.4
III	39	50	1.9
IV	85	64	9.2
V	90	69	16

The rock mass properties such as RQD, RMR, Q etc. were also determined from the above tests, borehole logs and field observations for each coal seam. These values are presented in Table 4.29.

The uniaxial compressive strength value was used in the empirical pillar strength calculations and the RMR values were used along with the other physico-mechanical properties for the numerical modelling.

4.3.2   Strategy for extraction

To protect the surface features, especially the flood protective bund and the river bank, it was necessary to design the web pillars for long-term stability. Considering the seams available for extraction and other geo-mining conditions the following points were taken into consideration when designing a suitable extraction method for the locked-up coal in the proposed blocks.

1   All the seams were to be extracted from bottom to top, starting from seam V and proceeding upward to seam I, by filling the pit with overburden rock which would act as a platform for the launch vehicle. Since the web pillars were to be designed for long-term stability, with $SF \geq 2.0$, from a stability point of view, the order of extraction and protection of overlying seams were not issues.

2   Web pillars were to be designed considering the highest depth at which they would be standing and the greatest mining height of that seam; this would correspond to the lowest safety factor throughout the length of pillars. The lowest safety factor thus designed would be ≥ 2.0 for long-term stability.

3   All the workable seams were separated by vertical partings of sufficient thickness (>9.0 m). The seams were therefore not contiguous, and they were designed separately. It had been previously found (CSIR-CIMFR, 2008) that the workings of one seam would not interact with the other if the vertical parting was more than 9.0 m. This also helped in maximising the recovery from each seam. However, inter-seam interaction was studied more precisely using numerical modelling before finalising the design.

4   Due to the existence of a major fault passing through the proposed block, the block had been sub-divided into two: block A and block B. A barrier of minimum 15 m thickness is proposed to be left against the fault in all the seams on both the sides, as shown in Figure 4.26.

5   The possible hole deviation for the ADDCAR system was considered as given earlier in Section 4.4.1. It was necessary that the average SF of the pillars was ≥ 2.0, and at the same time the minimum SF resulting due to possible hole deviation would not be lower than 1.9.

6   All the entries in all the targeted seams were limited to a distance from the mine boundary/river bank corresponding to $H \tan \theta$, where H is the seam depth at the end of the entries and θ is the angle of draw, which was taken as 25°. This was to prevent any collapse in the future, from reaching the river bank. This demarcation of a safe distance based on an angle of draw is given in the vertical section C–C′ shown in Figure 4.29.

7   As found from earlier studies (CSIR-CIMFR, 2009), when the slenderness ratio (W/h) of pillars goes below 1.0, the strength needs to be reduced by 20% to account for the shape factor corresponding to an equivalent square pillar. Therefore, whenever the W/h ratio came closer to 1.0 (less than 1.5), the pillar strength was reduced by 20%.

8   Web pillars were designed for each workable seam existing in block A and block B separately.

9   The lesson learned from the previous highwall mining site OCP-II were that many of the holes could not be penetrated to their designed hole depth due to unforeseen

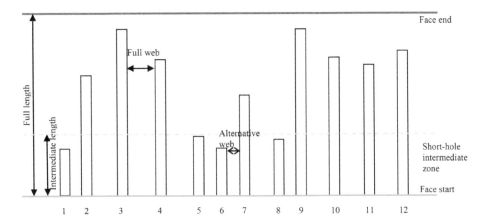

Figure 4.31 Concept of using a full-length web thickness and an alternative web thickness for short and incomplete holes.

geological disturbances such as local faults, stone bands occurring within the seam, abrupt changes in dip, seam thickness variations, etc. The conditions at MOCP also looked similar with a large number of faults. In certain blocks where the depth variation was high, when a hole was discarded midway, it was sometimes not necessary to leave the originally designed web pillar width for the next hole. To save some coal, a thinner web pillar could be left with the required safety factor for that depth. Considering the above aspect, for MOCP, in most of the blocks, an alternative web thickness was also suggested with a specified intermediate hole penetration length. If the previous hole fell short within the intermediate penetration length, the alternative web thickness was to be followed for the next hole. But if the hole was penetrated beyond this stipulated length, the original web thickness pertinent to the full hole length needed to be strictly followed for the next hole. The concept of alternative web thickness is given in Figure 4.31.

4.3.3   Highwall mining blocks

The highwall mining blocks, block A and block B, in each workable seam were demarcated in plans by taking into consideration the following aspects:

- the total area of highwall available for extraction is given in Figure 4.26;
- the incrops of top seams within the area. The workable seams incropping within the boundaries were seam IIIA in block A, seam II and seam I in block B. Seam II and seam I were absent in block A;
- a safe distance corresponding to the angle of draw from the river bank.

   Using the above details, the mineable regions in each workable seam for block A and block B were demarcated as given in Figures 4.32 to 4.36 for seams V to I, respectively. The end limit of the web cuts in each seam for both blocks are shown as green lines in these figures. The web cuts were strictly not supposed to cross the

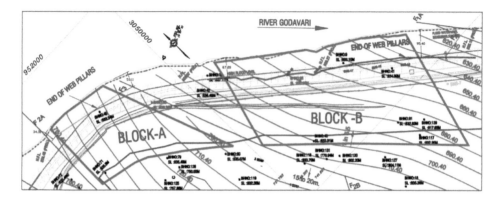

Figure 4.32   Highwall mining blocks in the horizon of seam V showing the limiting line of web cuts from the river bank.

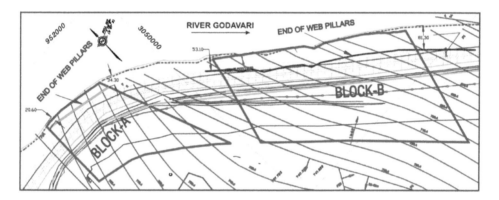

Figure 4.33   Highwall mining blocks in seam IV horizon showing the limiting line of web cuts from the river bank.

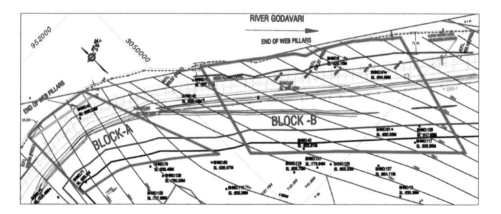

Figure 4.34   Highwall mining blocks in seam III horizon showing the limiting line of web cuts from the river bank.

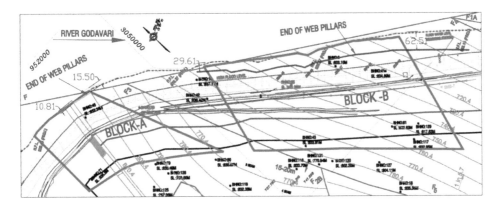

Figure 4.34a Highwall mining blocks in seam IIIA horizon showing the limiting line of web cuts from the river bank.

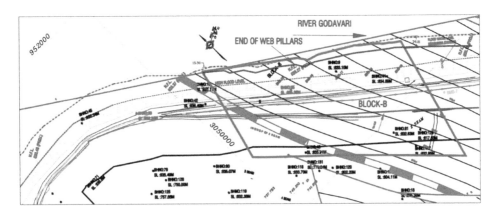

Figure 4.35 Highwall mining blocks in the horizon of seam II showing the limiting line of web cuts from the river bank.

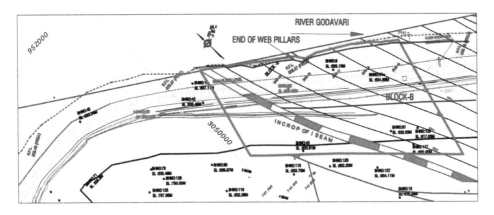

Figure 4.36 Highwall mining blocks in the horizon of seam I showing the limiting line of web cuts from the river.

Figure 4.37 Officials discussing on the strategy of extraction at the site.

limiting line. The safe distances calculated for each seam within the highwall mining blocks are also given at the end points of the blocks in Figures 4.32 to 4.36.

4.3.4   Empirical web pillar design

Web pillars were designed for all the seams separately for each block. The exposed coal seams at the highwall, namely seam I, seam II, seam III, seam IIIA, seam IV and seam V, occurred in descending order. The methodology adopted for the empirical design included:

- a general strategy for extraction of individual seams such as order of extraction, method of workings, etc. (Figure 4.37);
- design of web pillars in specific sites demarcated at block A at the northern side and block B towards the east side of the pit;
- empirical and numerical modelling studies undertaken for pillar and entry stability;
- design of controlled blasting parameters for the open-pit bench blasting for the safety and stability of the highwall mining due to ground vibration and flyrock;
- recommendations for highwall stability.

4.3.4.1   Design for seam V

Seam V being the bottom-most seam in block A and block B, was to be extracted first after the final highwall was reached and the pit bottom was de-coaled.

The following points were kept in mind while designing web pillars for these two blocks:

- extraction to full seam thickness, as possible;
- long-term stability of the web pillars (safety factor ≥ 2.0);

Table 4.30 Web pillar design for seam V.

Block A – seam V					Block B – seam V				
Full-length holes		Intermediate length holes			Full-length holes		Intermediate length holes		
Max. length	Web width	Int. length	Alt. web	% ext.	Avg. length	Web width	Int. length	Alt. web	% ext.
230 m	3.2 m	70 m	2.4 m	52.2%	240 m	5.8 m	70 m	3.1 m	37.6%

- if a hole fell short within the intermediate penetration length, an alternative web thickness was specified for block A and block B to be followed for the next hole, while maintaining SF ≥ 2.0 throughout the length;
- in block A the extraction thickness would decrease by 0.5 m than the vertical seam thickness since the machine would be operating roughly along the strike (apparent dip);
- web pillar design considered maximum possible extraction height and maximum depth at which they would exist.

Using the above strategy, web pillar sizes given in Table 4.30 were designed for seam V.

4.3.4.2 Design for seam IV

Seam IV would be approached after the extraction of seam V and dumping the pit with overburden material to form a platform. Seam IV had a maximum vertical thickness of 4.5 m, and therefore could be extracted in one pass. The following points were taken into account while designing web pillars in seam IV:

- Seam IV had to be extracted to the maximum possible mining height. Due to the apparent dip, in block A, the extraction height would be about 0.5 m lower than the full seam height.
- Alternative web thicknesses were specified for intermediate hole lengths for block A and block B.
- For both the blocks, web pillars were designed so as to have long-term stability (SF ≥ 2.0) everywhere.
- Web pillars were designed considering the maximum possible extraction height in that region and the maximum depth of cover.

Taking the above aspects into account, the web pillars were designed as per Table 4.31.

4.3.4.3 Design for seam III

Seam III was the thickest seam existing in the Medapalli OCP block, with a vertical thickness of up to 6.5 m. Therefore, some parts of this seam could be extracted using two-pass cutting to enable full seam thickness extraction. The web pillars were

Table 4.31 Web pillar design for seam IV.

Block A – seam IV					Block B – seam IV				
Full-length holes		Intermediate length holes			Full-length holes		Intermediate length holes		
Max. length	Web width	Int. length	Alt. web	% ext.	Avg. length	Web width	Int. length	Alt. web	% ext.
220 m	5.7 m	70 m	4.0 m	38%	225 m	9.9 m	70 m	6.0 m	26.1%

Table 4.32 Web pillar design for seam III.

Block A – seam III					Block B – seam III				
Full-length holes		Intermediate length holes			Full-length holes		Intermediate length holes		
Max. length	Web width	Int. length	Alt. web	% ext.	Avg. length	Web width	Int. length	Alt. web	% ext.
200 m	6.0 m	60 m	4.8 m	36.8%	215 m	11.1 m	60 m	6.5 m	24%

designed for the maximum possible thickness. The following points were taken into consideration for the empirical design of seam III extraction:

- Seam III had to be extracted to the maximum possible seam height; two-pass cutting would be employed wherever the thickness exceeded 4.0 m.
- Alternative web thicknesses were specified for intermediate hole lengths for block A and block B.
- Web pillars would have safety factors ≥ 2.0 at all instances, even at the maximum extraction height and the maximum depth of cover.

The web pillar design for seam III in block A and block B is given in Table 4.32.

4.3.4.4 Design for seam IIIA

Seam IIIA was a thin one and occurred as the top-most in block A, but was overlaid by seam II and seam I in block B. All extraction cuts were designed with full-length penetration in block A and block B. Following were the salient points of extraction design for seam IIIA:

- Seam IIIA had to be extracted to a full seam vertical thickness of about 2.0 m in block B and about 1.5 m in block A.
- Alternative web thicknesses were specified for intermediate hole lengths for block B alone; there was no scope for reducing the web pillar thickness in block A.
- Web pillars would have a safety factor of ≥ 2.0 at all instances.

Table 4.33 Web pillar design for seam IIIA.

| Block A – seam IIIA | | | Block B – seam IIIA | | | | |
| Full-length holes | | | Full-length holes | | Intermediate length holes | | |
Max. length	Web width	% ext.	Avg. length	Web width	Int. length	Alt. web	% ext.
155 m	1.6 m	68.6%	185 m	3.8 m	60 m	2.2 m	47.9%

Table 4.34 Web pillar design for seam II.

| Block B – seam II | | | | |
| Full-length holes | | Intermediate length holes | | |
Avg. length	Web width	Int. length	Alt. web	% ext.
145 m	3.7 m	60 m	3.0 m	48.6%

Table 4.33 shows the design parameters for web pillars in block A and block B for seam IIIA.

4.3.4.5 Design for seam II

Seam II outcropped at the block B region, before the fault separating the two blocks, and was therefore non-existent in block A. Web thicknesses for full-length and intermediate extraction hole lengths were designed for block B for the full vertical seam thickness. The web pillars would have a safety factor ≥ 2.0 throughout the block. Alternative web thicknesses were specified if the holes stopped before the intermediate length. The web pillar design parameters for seam II in block B are given in Table 4.34.

4.3.4.6 Design for seam I

This was the top-most seam in the mine boundary of Medapalli OCP and it existed in a small patch in block B alone. The seam had a good vertical thickness, with a maximum of about 4.5 m. Here also web thicknesses for full-length and intermediate extraction hole lengths were designed for block B for the full vertical seam thickness. The web pillars would have a safety factor ≥ 2.0 throughout the block. The web pillar design parameters for seam II in block B are given in Table 4.35. A percentage of recovery of about 48.6% was designed for seam I. The above empirical design had been done purely on the basis of pillar stability, as observed for the past several decades in Indian coal mining scenario. However, entry roof stability and vertical parting stability also might influence the pillar strength in the long run. For example, a very weak roof may

Table 4.35 Web pillar design for seam I.

Block B – seam I				
Full-length holes		Intermediate length holes		
Avg. length	Web width	Int. length	Alt. web	% ext.
125 m	3.7 m	50 m	3.2 m	48.6%

result in an increase in the effective working height and may thus reduce the pillar strength in the long run. Further, if two coal seams are occurring contiguously, with very less intervening parting, the parting may eventually fall and thus increase the effective pillar height. These aspects can be studied only by numerical modelling since there are no empirical equations available for such special cases.

4.3.5 Numerical modelling for analysis of overall stability

The overall stability, including the roof stability of entries, inter-seam interaction parting stability and pillar stability were analysed using elastic numerical modelling studies. Two models were constructed, one representing block A and the other for block B. These models took into account the web cuts existing at the highest depth in each block, thus simulating the highest stress state. These models had geo-mining parameters taken from the design Tables 4.30–4.35 for seams V to I, respectively.

Numerical modelling was conducted using *FLAC3D*. The safety factor block contours presented were evaluated using a programming environment (*FISH*) of the software. The modelling and theory behind the local safety factor evaluations are presented in the following sections.

4.3.5.1 Rock properties used for modelling

Physico-mechanical properties used in the elastic models are those tested previously by the NIRM and provided by AMT and SCCL, as given in Table 4.28. The RMR values taken for coal and parting rock were the average values.

4.3.5.2 Modelling procedure

Numerical modelling studies had been done in the following stages:

Stage 1: The 2D models were constructed for block A and block B separately considering the plane-strain scenario. The virgin stage was modelled, loaded with the elastic properties of the different rock types and *in situ* stresses as given for strain-softening modelling.
Stage 2: Roadways in each seam were excavated and the model was run to equilibrium.
Stage 3: The *FISH* program was run using the output of the numerical model along with strength parameters of different rock types to analyse the roadway, parting and pillar stability in terms of safety factor contours.

Block contours of safety factor

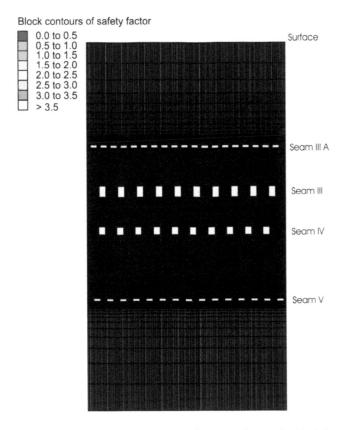

Figure 4.38 Safety factor contours in the entire domain for block A.

4.3.5.3 Results

The results of the numerical modelling were analysed in terms of local safety factor contours.

Block A

Figure 4.38 shows the block contours of the safety factor throughout the domain modelled for block A. A closer view of the safety factor contours in the top-most seam IIIA and the rock mass between the surface and seam IIIA is given in Figure 4.39(a). The safety factor contours in the vicinity of seams III and IV are given in Figure 4.39(b). The stability of the intervening parting between seams III and IV can also be seen from Figure 4.39(b). The stability of the bottom-most seam V is shown separately in Figure 4.39(c).

The roofs of all the workings had satisfactory stability from a short term point of view. The roof stability was needed only for a day when the CHM would be operating in a web cut traversing up to 300 m. The local safety factors shown in the roof of the workings were mostly greater than 1.0.

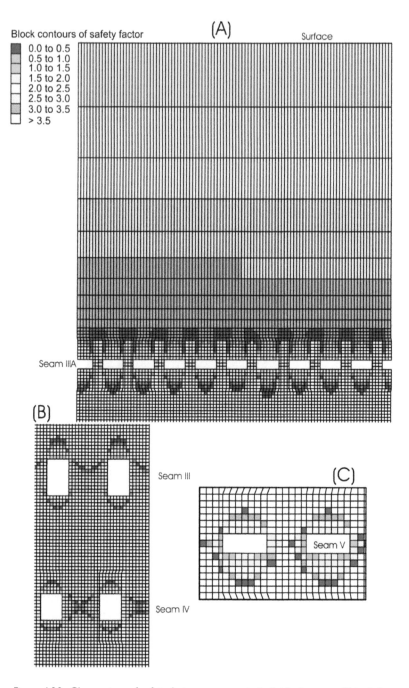

Figure 4.39 Closer view of safety factor contours in individual seams of block A.

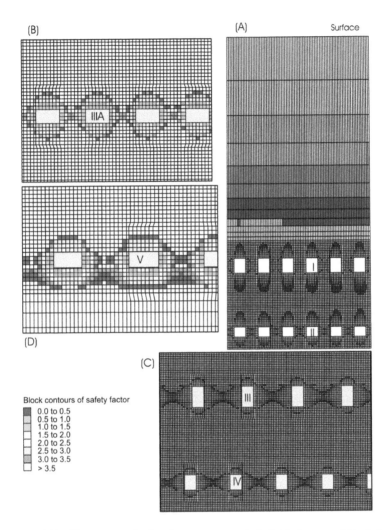

Figure 4.41 Closer view of safety factor contours in individual seams of block A.

4.3.7 Impacts of surrounding blasting

Blasting activities close to the highwall mining at Medapalli Opencast Project might affect the stability of the highwall mining entries, web pillars as well as the highwall slopes. Ground vibrations generated from day-to-day blasting operations nearby the highwall mining entries could cause premature roof and side collapse, thereby possibly trapping the continuous miner inside the mine entries. Flyrock generated from blasting could also damage the highwall mining equipment and result in safety of miners.

4.3.7.1 Site inspection

The site inspection was carried out by the CSIR-CIMFR team to visualise the present working conditions as well as the highwall mining site of Medapalli OCP. The nearest

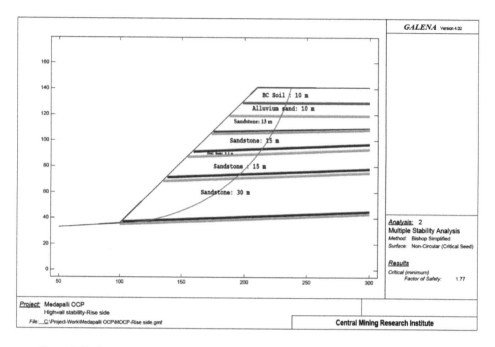

Figure 4.42 Stability analysis of the rise/strike side (block A) highwall along section C–C'.

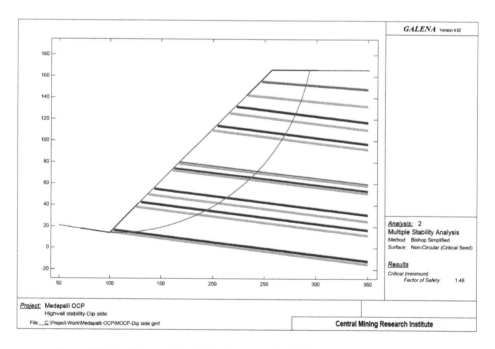

Figure 4.43 Stability analysis of the dip side highwall (block B) along section B–B'.

Table 4.36 Details of various seams for proposed highwall mining at Medapalli OCP.

Parameter	Minimum (m)	Maximum (m)
Seam I cover	26	71
Seam I thickness	2.5	4.5
Seam II cover	25	90
Seam II thickness	2.5	3.0
Seam IIIA cover	27	141
Seam IIIA thickness	1.5	2.0
Seam III cover	30	170
Seam III thickness	5.0	7.0
Seam IV cover	44	192
Seam IV thickness	3.0	5.0
Seam V cover	83	224
Seam V thickness	1.0	2.0

villages from the present blasting site of the mine was situated at around 1.8 km (Pamulapeta village & Lingapur village). The highwall mining site was located at around 600 m from the present blasting site. Around 10 tonnes of explosives were being utilised at Medapalli OCP in one round of blasting to meet the annual coal production of 3 million tonnes. The bench height was 9–10 m whereas the depth of holes varied between 5–6 m. The hole diameter was 150 mm. The number of holes being fired in a blasting round varied between 100 and 300 whereas in each hole around 80–90 kg explosives were being used. The burden and spacing used in the blasts were 4.0 m and 5.5 m respectively.

4.3.7.2 Assessment of safe value of PPV

As per the DGMS Technical Circular (DGMS, 2007), the permissible limit of peak particle velocity (Tables 4.18 and 4.19) for the stability of roof and web pillars of highwall mining depends on the RMR values of the immediate roof. The details of various seams are given in Table 4.36. In total, six seams, I, II, IIIA, III, IV and V, were targeted for highwall mining at the Medapalli OCP. The nature of the roofs and their respective RMR values for different seams are given in Table 4.37. For seam III, two RMR ratings had been provided by the mine management based on past studies. One RMR rating was corresponding to a 4.6 m mining height (assuming a coal roof) while the other was corresponding to a 7 m mining height (assuming a rock roof).

The lowest RMR value from the different rock/coal roofs was 45 for seam III. Seams II and IV had the similar values of RMR (i.e. 64 each). Therefore, considering the lowest value of RMR as the safest criterion, the threshold value of peak particle velocity for the safety of highwall mining entries came to 90 mm/s (as per DGMS, 2007).

Based on the generalised criteria proposed by Bauer and Calder (1971) (Table 4.20) for the damage level of particle velocity on rock mass and slopes, the safe value of the threshold PPV (i.e. 90 mm/s) was suggested for the safety of the roofs of the highwall entries, which would also be applicable for highwall slopes. The highwall slopes would also have less damage impact due to ground vibration in comparison to the roofs of the galleries.

Table 4.37 Nature of roof and corresponding RMR values of different seams.

Name of Seam	Nature of the immediate roof	RMR value
I	Sandstone	57
II	Medium-coarse grained sandstone	64
IIIA	Fine-coarse grained sandstone	55
III	Shale-coal roof (up to 4.6 m) sandstone	45
		50
IV	Sandstone	64
V	Sandstone	69

4.3.7.3 Assessment of safe blasting parameters

The CSIR-CIMFR team carried out a number of scientific investigations at the opencast mines of Singareni Collieries Company Limited (SCCL) which have similar geo-mining conditions as those of Medapalli OCP. The blast design parameters experimented with at various opencast mines of SCCL had been used to analyse the outcome of the blasting operations being used at Medapalli OCP.

The ground vibration data recorded during the field investigations by the department at various opencast mines of SCCL under similar geo-mining conditions were grouped together for a rigorous statistical analysis. Based on the analysis, an empirical equation had been developed for prediction of safe maximum explosive weight per delay, correlating the distance of vibration measuring transducers from the blast face (D in m), the maximum explosive weight per delay (Q_{max} in kg) and recorded peak particle velocity (V in mm/s). The regression plot of the vibration data is given in Figure 4.44. The empirical equation for the prediction of the maximum charge per delay to be fired in a blasting round for the safety of highwall entries and slopes at Medapalli OCP was suggested as given below:

$$V = 1077[D/\sqrt{Q_{max}}]^{-1.37} \tag{4.14}$$
Coefficient of determination $= 0.837$
Standard deviation $= 0.233$

This equation could be used to compute the safe maximum explosive weight to be detonated in a delay for a specific distance between the blasthole and the highwall entries. Bearing in mind the safe peak particle velocity values for the total safety of highwall mining entries, the safe levels of maximum explosive weight per delay had been calculated for various distances, using Equation (4.14) and are given in Table 4.38. The same is also represented in Figure 4.45. However, for better safety, it was advised to contain the total firing time (i.e. difference between the firing times of the last hole to the first hole) of the blasting round within 1000 millisecond i.e. 1 second.

4.3.8 Surface ground movement monitoring

Surface ground movement monitoring is an imperative part of executing a safe and smooth highwall mining operation. Field measurement is also beneficial to validate

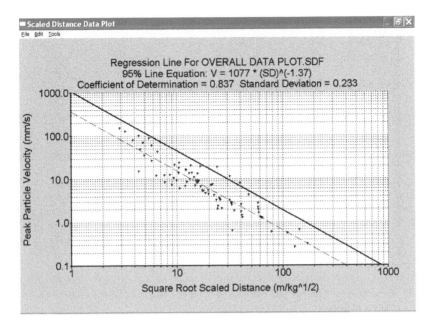

Figure 4.44 Regression plot of vibration data recorded at various opencast mines of SCCL.

Table 4.38 Safe explosive charge per delay at various distances from the blasting site.

Distance from the blasting face [m]	Suggested maximum charge per delay [kg]
50	67.6
100	270
200	1083
300	2434
400	4328
500	6762
600	9738
700	13254
800	17312
900	21910
1000	27050

the predicted values as well as for further refining the input parameters, based on the generated field data, for accurate prediction.

The surface subsidence and slope stability of the highwall had to be monitored at regular intervals during the mining operations. Subsidence measurements were to be continued for at least 6 months after mining to evaluate the stability of the ground subsequent to mining.

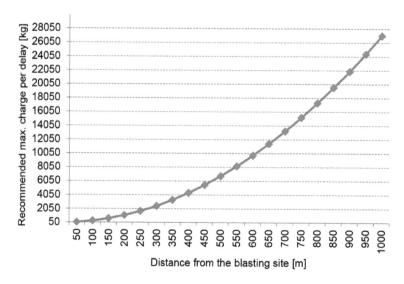

Figure 4.45   Recommended charge per delay to be fired in a round of blast for the safety of highwall mining entries.

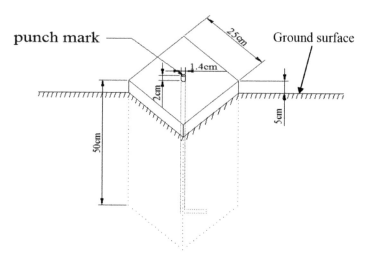

Figure 4.46   Design of subsidence pillar.

The slope stability of the highwall could be monitored all throughout the wall whenever feasible to access the benches, otherwise measurement would be carried out at least at the top edge of the highwall. The layout of the subsidence monitoring stations would cover the complete working area (inclusive of angle of draw from either edge of the bottom-most seam). The monitoring stations had to be fixed at an interval of 5 m as per the designed layout. The dimensions of the monitoring station are given in Figure 4.46. The reinforced cement concretepillar would be projected at least 5 cm

Figure 4.47 Auto level with micro drum.

Figure 4.48 Total station.

above the ground and the centrally fixed 1.4 cm diameter MS rod had to be projected out by at least 2 cm.

4.3.8.1 Methodology

Auto level (Figure 4.47) and total station (Figure 4.48) were recommended to carry out the measurement of the subsidence and stability of the highwall.

Vertical ground movement (subsidence) would be monitored by auto level, the least count of the instrument being 0.1 mm. The accuracy of the instrument had been

±1 mm. Though, subsidence measurement could be carried out by total station, its accuracy decreases as the distance between target and instrument increases. Auto level is the most preferable instrument for subsidence measurement especially in sensitive and critical areas. It was therefore chosen.

For computing the horizontal strain (compressive and tensile), slope of subsidence and lateral displacement of ground, total station was chosen. The least count in vertical and horizontal measurements had been 1 mm and the accuracy of the instrument was $+ (2\,mm + 2\,ppm \times D)$ where D is the distance in mm.

4.3.9 Summary of design and recommendations

From the empirical design and numerical modelling the following conclusions were drawn:

1   The entire region targeted for highwall mining was divided into two blocks, block A towards north and block B towards east of the Medapalli OC Project (Figure 4.26).

2   The highwall mining blocks were demarcated at each workable seam horizon, as shown in Figures 4.32 through 4.36, by taking a safe distance from the river bank, equal to $H \tan \theta$, where H is the depth of web cut towards the end and θ is angle of draw, which is 25°.

3   As observed at OCP-II, due to unforeseen geological conditions, many holes could not be penetrated much to their designed hole depth. To minimise the coal loss during identical scenarios, an alternative web thickness was also suggested with a specified intermediate hole penetration length for each seam and block. When the holes fell short of the specified intermediate length, the alternative web thickness had to be followed.

4   All workable seams in block A, seam V, seam IV, seam III and seam IIIA, were to be extracted to full seam thickness. Seam III, in places, required two-pass cutting due to its higher vertical thickness.

   - The recommended web pillar thicknesses for seam V, seam IV, seam III and seam IIIA for full-length holes were 3.2 m, 5.7 m, 6.0 m and 1.6 m, respectively.
   - The alternative web pillar thicknesses, if the penetration stopped before the intermediate hole lengths for seam V, seam IV and seam III, were 2.4 m, 4.0 m and 4.8 m, respectively. For the top-most seam IIIA no such alternative web thickness was specified in block A. The intermediate hole lengths specified for seam V, seam IV and seam III were 70 m, 70 m and 60 m, respectively. The above design would result in a recovery of 52.2%, 38%, 36.8% and 68.6% for seams V, IV, III and IIIA, respectively.

5   All workable seams in block B, seams V, IV, III, IIIA, II and I, could be extracted to their possible full seam thickness. In this block also seam III required double-pass cutting in places to achieve full seam extraction.

   - The recommended web pillar thickness for seam V, seam IV, seam III, seam IIIA, seam II and seam I for full-length holes were 5.8 m, 9.9 m, 11.1 m, 3.8 m, 3.7 m and 3.7 m, respectively.
   - The alternative web pillar thickness, if the penetration stopped before the intermediate hole lengths for seam V, seam IV, seam III, seam IIIA, seam II and

seam I, were 3.1 m, 6.0 m, 6.5 m, 2.2 m, 3.0 m and 3.2 m, respectively. The intermediate hole lengths specified for seam V, seam IV, seam III, seam IIIA, seam II and seam I were 70 m, 70 m, 60 m, 60 m, 60 m and 50 m, respectively.

- The above design would result in a recovery of 37.6%, 26.1%, 24%, 47.9%, 48.6%, and 48.6% for seam V, seam IV, seam III, seam IIIA, seam II and seam I, respectively.

6   The numerical modelling studies revealed no significant inter-seam interaction, superior stability of the intervening parting, satisfactory roof stability and good web pillar stability at all instances of extractions in all seams modelled for block A and block B.

7   The highwall slopes on the dip side and rise/strike side of the MOCP were fairly stable for an overall slope angle of 44° with factors of safety ranging between 1.4 and 1.75 as indicated from the stability analysis results. The recommendations given in Section 4.2.6 for highwall stabilisation needed to be followed.

8   Blasting activities close to the proposed highwall mining at Medapalli OCP could affect the stability of highwall mining entries, web pillars as well as the highwall slope. As such, a controlled blasting operation had to be carried out at such locations.

9   The lowest RMR value of the rock/coal roof found in the working area was 45 (shale-coal roof of seam III). Seams II and IV had similar values of RMR (i.e. 64 each). Therefore, considering the lowest value of RMR as the safest criterion, the threshold value of peak particle velocity for the safety of highwall mining entries came to 90 mm/s (DGMS, 2007).

10   Based on the generalised criteria proposed by Bauer and Calder (1971) (Table 4.20) for the damage level of particle velocity on rock mass and slopes, the threshold value of PPV (i.e. 90 mm/s) was suggested for the safety of the roofs of the highwall entries, which would also be applicable for the highwall slopes. The highwall slopes would also have less damage impact due to ground vibration in comparison to the roofs of the galleries.

11   The propagation equation for the prediction of safe maximum explosive weight per delay in a blasting round at Medapalli OCP had been established and is given as Equation (4.14). Bearing in mind the safe peak particle velocity values for the total safety of highwall mining entries, the safe levels of maximum explosive weight per delay had been calculated for various distances of concern, using Equation (4.14) and are given in Table 4.38. More experimental blasts and monitoring of vibration data near highwall entries could make empirical models further refined.

4.4 SITE C

HIGHWALL MINING AT QUARRY SEB (AND QUARRY AB), WEST BOKARO OF M/S TATA STEEL LIMITED

The Tata Steel Limited (TSL), West Bokaro Division proposed to extract the locked-up coal in the final pit slope of Quarry SEB and that lay below Banji village in Quarry AB at West Bokaro Group of Collieries using the Continuous Highwall Mining (CHM)

Figure 4.49 View of 15 working benches of both coal and OB of Quarry SE of TSL, West Bokaro.

system. In that context, CSIR-CIMFR carried out the feasibility study on the design and safety aspects of highwall mining for implementation.

At Quarry SEB seams V, VI, VII, VIII, IX, X Lower, X Upper and XI existed in ascending order and were exposed on the highwall towards the eastern side of the quarry. At Quarry AB, the coal seams left below Banji village and those considered feasible for extraction from the final highwall abutting the village boundary included seams V, VI, VII and VIII, occurring in ascending order. All the seams were gently dipping, with 3–7° inclination and occurred at a depth of 16–150 m within the proposed highwall mining sites.

At Quarry SEB, no significant surface features existed, whereas Banji village has been densely populated with residential houses, to be protected from surface subsidence on a long-term basis. The final pit slope in both the quarries had been designed by CSIR-CIMFR (CSIR-CIMFR, 2000; 2008b). The highwall would have an overall angle of 47° from the horizontal at both the quarries. This angle was considered as the final pit slope angle for the design of the highwall mining. Photographic views of the final highwall and surface features of Quarry SEB are shown in Figures 4.49 and 4.50.

In this study, the feasibility of extracting coal from the above targeted seams at specific locations in both the quarries was analysed and schemes for such extractions were designed for individual seams maximising coal recovery with required safety factors for the protection of surface and sub-surface properties from subsidence.

The plan views of the areas to be extracted by highwall mining at Quarry SEB and below Banji village at Quarry AB are given in Figure 4.51 and Figure 4.52, respectively.

The study covered the following areas:

- A general strategy for extraction of individual seams such as order of extraction, requirement of vertical superposition of workings, etc.
- Design of web pillars in specific sites demarcated at Quarry SEB for individual seams with a short-term stability point of view

Figure 4.50 View of road and dwellings of Banji village below which highwall mining has been proposed near Quarry AB.

- Design of web pillars in individual seams below Banji village with long-term stability point of view
- Details of empirical and numerical modelling studies undertaken for pillar and entry stability

4.4.1 Assumptions about the machine operating parameters

Prior to the web pillar design at West Bokaro, the following assumptions were made regarding the highwall mining operations:

1   Out of the continuous highwall mining equipment in use world-wide, most of the latest equipment operate with 3.5 m wide web cuts. The web pillars were designed for 3.5 m wide cuts.
2   A maximum penetration depth of 500 m was assumed.
3   The CHM machines to be used would be able to cut up to 4.5 m height in a single pass and up to 8.0 m in a double pass.
4   An important aspect in web pillar design is the accuracy of the cuts, which determines the minimum and maximum size of the resulting web pillars. The required strengths and safety factors of pillars are to be decided based on the maximum deviation of the roadways and thus the resulting minimum thicknesses of the web pillars. Manufacturers of the CHM equipment usually specify a cutting accuracy in terms of the penetration depth. For the top seams to be extracted below Banji village, the web pillar might be of a width of 2–3 m with a proposed penetration of 500 m. In such a case, poor accuracy might result in a very narrow pillar or even a crossing of entries, leading to surface subsidence. Therefore it was necessary that the CHM to be employed would have a maximum deviation within certain limits. For the current design an accuracy of 0.2% i.e. a deviation of 20 cm per 100 m penetration depth was assumed. The resulting web pillar, after accounting for the deviation, would have a width $W_d = W \pm 0.4 * L/100$, where the variation in pillar width was calculated twice as that of the entry deviation, and L was the maximum penetration depth.

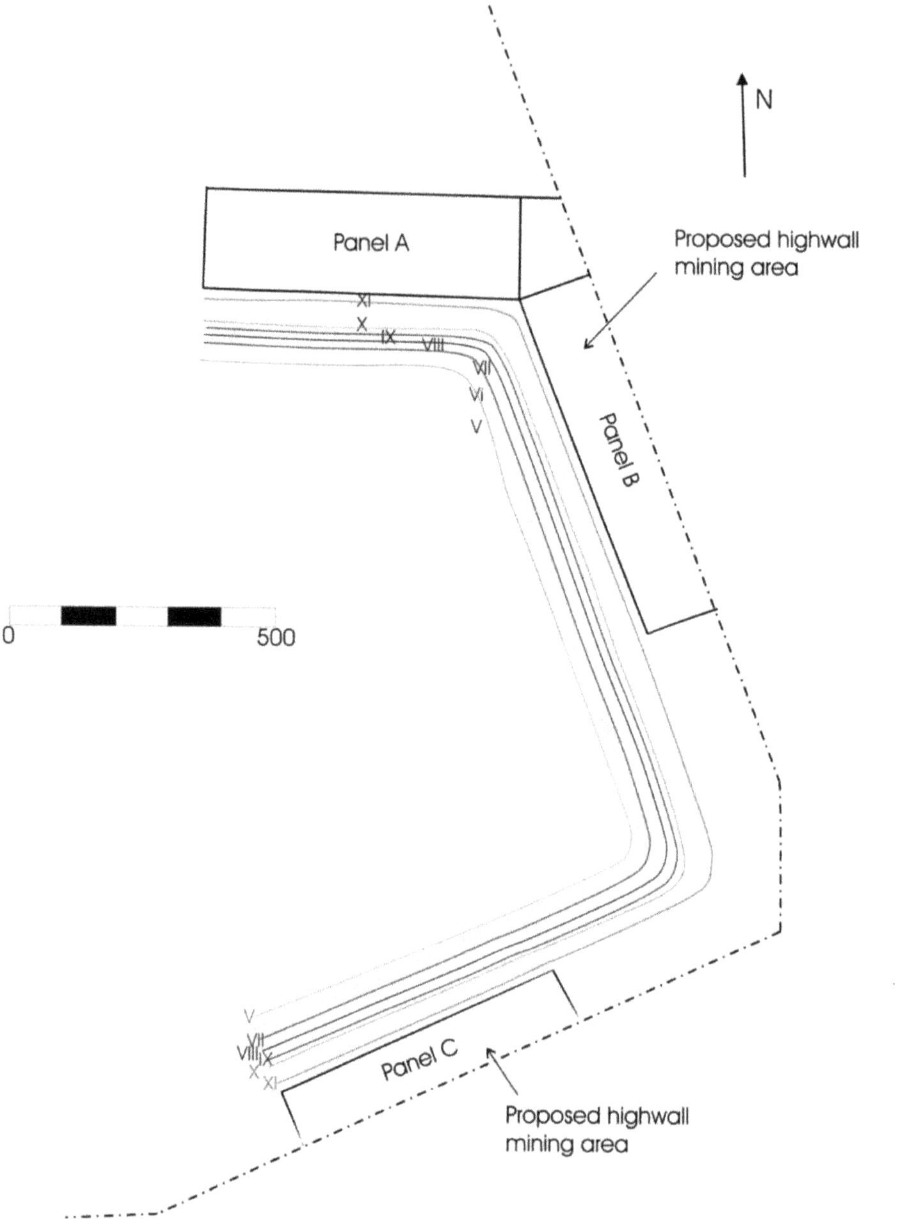

Figure 4.51 Proposed highwall mining areas in Quarry SEB and the seams exposed on the final pit slope.

4.4.2 Geo-technical information

Pertinent information regarding rock and rock mass properties of the targeted coal seams and intervening parting were provided by M/s Tata Steel Limited (TSL).

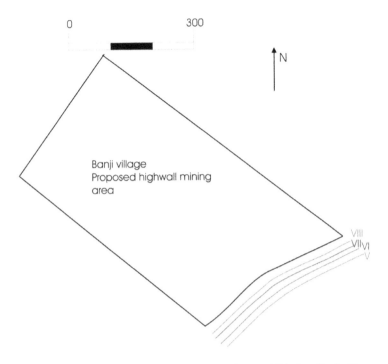

Figure 4.52 Proposed highwall mining area in Quarry AB below Banji village.

Table 4.39 Seam details at highwall mining area in Quarry SEB.

Seam	Approx. depth (m)	Average thickness (m)	Parting (m)
XI	20	2.7	–
X U	75	1.4	40–50 with seam XI
X L	77	1.2	0–1 with seam X U
IX	83	2.0	7–20 with seam X L
VIII	87	4.2	2–4 with seam IX
VII	105	8.0	11–30 with seam VIII
VI	125	4.0	10–20 with seam VII
V	143	4.0	5–20 with seam VI

The crushing strength of 1 inch cube coal samples for a few seams had been tested in the laboratory by CSIR-CIMFR. The strength of 1 inch cubes is an integral part of the CSIR-CIMFR empirical pillar strength equation. The crushing strength of coal seams from earlier tests at CSIR-CIMFR had been found to be about 20–24 m MPa. A minimum value of 20 MPa was used for the design.

A summary of the seams proposed to be extracted at Quarry SEB is given in Table 4.39.

Table 4.40 Seam details at Quarry AB below Banji village.

Seam	Approximate depth (m)	Average thickness (m)	Parting (m)
X	14	1.5	–
IX	16	1.0	0–1 with seam X
VIII	20	3.0	3.0 with seam IX
VII	30	8.0	7–13 with seam VIII
VI	60	4.5	14–25 with seam VII
V	75	3.0	8–15 with seam VI

As evident from the borehole sections provided by TSL, the top-most workable seam at Quarry SEB would be seam XI occurring at a depth of more than 15 m from the surface. The surface cover at all the sites comprised mostly of hard strata and therefore seam XI was considered to be workable. Similarly, the top-most workable seam below Banji village was seam VIII occurring below about 20 m depth of cover. Due to the scanty information available from the boreholes, TSL was advised to accurately assess the hardcover for the top seams prior to extracting by highwall mining.

Seams occurring above seam VIII at Banji village were IX and X; these were at less than 15 m depth of cover in places, and therefore as per DGMS guidelines, were not to be extracted to avoid the occurrence of pot-holing below the village. The targeted seams in Quarry AB below Banji village are given in Table 4.40.

At West Bokaro, the intervening parting comprised mostly of sandstone with some layers of shale and mudstones. At some places of Quarry AB, seam VII had been developed by the underground bord and pillar method. The occurrence of some faults had also been seen at both the pits. From the boreholes drilled and investigated by TSL, the rock properties used as input parameters in numerical modelling for the coal seams and partings are as given in Table 4.41.

4.4.3 Strategy for extraction

Depth variations of the seams were not very significant at all the locations due to the flat gradient and therefore an approximate maximum depth, as could be inferred from the boreholes, was used for the pillar design.

Due to the discrete layout of the extraction areas in Quarry SEB, the area proposed for highwall mining was demarcated into three panels: panel A, panel B and panel C as shown in Figure 4.51. However, due to the very little variation in the depth and thickness of the individual seams, the web pillar design was done for all panels based on the maximum values.

At Quarry SEB, the final pit slope had already been reached towards the south-eastern part of the pit, indicating the end of all possible opencast mining operations. Further benching was not possible in the 47° pit slope and hence the highwall mining could only progress upward from the bottom seam V to the top-most seam XI at that part of the pit. After the extraction of seam V it would be necessary to fill the pit bottom until the horizon of seam VI, forming a platform for its extraction, and so on

Table 4.41 Rock and rock mass properties used for empirical and numerical studies.

Strata	σ_c (MPa)	E (GPa)	ν	Density (kg/m³)	RMR
XI roof	35.5	10	0.25	2500	55
seam XI	20	2	0.25	1500	55
XI-XU parting	55	10	0.25	2500	65
seam XU	20	2	0.25	1500	55
XU-XL parting	49	10	0.25	2500	55
seam XL	20	2	0.25	1500	55
XL-IX parting	39.5	10	0.25	2500	65
seam IX	20	2	0.25	1500	55
IX-VIII parting	39.5	10	0.25	2500	55
seam VIII	20	2	0.25	1500	55
VIII-VII parting	35	10	0.25	2500	55
seam VII	20	2	0.25	1500	55
VII-VI parting	31	10	0.25	2500	55
seam VI	20	2	0.25	1500	55
VI-V parting	30.7	10	0.25	2500	60
seam V	20	2	0.25	1500	55

and so forth. For extraction of seams in ascending order it was necessary that the left-out web pillars would ensure the protection of the upper seams for the period they were required. Therefore, web pillars with short or medium-term stability, with a minimum safety factor of 1.5, were designed for all the proposed areas at Quarry SEB.

The area below Banji village in Quarry AB would be extracted as one panel. All the web pillars were to be designed for long-term stability, with a minimum safety factor of 2.0, and the sequence of extraction would not be of any significance. When the benches were available, the mining sequence might be descending. However, an ascending sequence as suggested for Quarry SEB could also be followed for this site as well, since there would not be any change in the dimensions of the resulting web pillars. The occurrence of surface and sub-surface subsidence needed to be completely avoided for the work below Banji village.

At Quarry SEB, seams X's upper section (X Upper) and seam X's lower section (X Lower) occurred at a vertical parting of less than 2.0 m. In some places they occurred as one seam. The following aspects needed to be taken into account for the extraction of seam X:

1 The web cuts in X's upper section needed to be vertically superposed with those of X's lower section.
2 The upper section needed to be extracted first followed by the lower section. This was due to the insufficient and inconsistent thickness of the intervening parting for supporting the continuous miner.
3 While designing web pillars, a combined extraction height of about 4.0 m was considered, including 1.2 m for XU, 1.4 m for XL and a 1.5 m parting. This would incorporate the overall pillar strength scenario in the event of the parting collapsing.

The parting between seam XL and seam IX varied between 7–20 m. Numerical modelling studies were proposed to evaluate the strength of this parting where it would be the minimum for assessing the requirement of vertical superposition of seam IX with seam X's workings.

Seams VIII and IX occurred at a vertical parting of 2–4 m and therefore were designed to be extracted with columnised pillars. However, the stability of the parting was to be assessed prior to finalising the pillar width for the individual seams. In this case, it was also advisable to extract the top seam IX followed by seam VIII.

At Quarry AB below Banji village, all the seams occurred at a considerable vertical parting with each other and the web pillars were designed individually. When the web pillars were designed for a minimum safety factor of 1.5, there would not be any requirement for barrier pillars or the omission of holes within an extraction area.

4.4.4 Verification of equivalent width for long pillars

The traditional empirical pillar design concepts, such as the ones mentioned above, had been developed from past case examples of square or rectangular pillars. In high-wall mining, long pillars without inter-connections are made by CHM. The concept of equivalent width $W_e = 4A/P$ (A: area, P: perimeter) is generally followed for irregular-shaped and rectangular pillars of aspect ratios more than 1 (Wagner, 1974). While applying $W_2 \gg W_1 = W$ for a long pillar, the equivalent width W_e mathematically becomes $2W$. Researchers in the past, such as Wagner (1974) and Galvin (1999), had conducted experimental studies to verify the above mathematical concept of equivalent width. Galvin (1999) suggested that the equivalent width should be applied only for moderate W/h ratios.

To verify the above equivalent width concept, elasto-plastic numerical modelling was conducted for a plane-strain scenario representing an array of infinitely long pillars of width 5 m, at a depth of 50 m with 3.5 m wide roadway; the extraction height was taken as 2.0 m, 4.0 m, 6.0 m and 8.0 m. An equivalent square pillar situation was modelled for 10 m × 10 m square pillar array keeping all other parameters the same as the previous set. For the long pillar configuration, two vertical planes of symmetry were taken, one passing through the centre of the roadway and the other through the centre of the long pillar. An equivalent width square pillar array was simulated by taking only a quarter portion of the pillar, bound by four vertical planes of symmetry, two passing through the centre of the pillar and the other two passing through the centre of the adjacent roadways.

4.4.5 Strain-softening modelling in FLAC3D

The *FLAC3D* software provides for elasto-plastic analysis of rock excavations with a strain-softening material model using the linear Mohr-Coulomb failure criterion. The strength and elastic constants necessary for numerical modelling using *FLAC3D* in the strain-softening mode are:

- elastic constants,
- peak and residual shear strength and its variation with the shear strain,
- peak and residual angle of internal friction and its variation with the shear strain, and
- angle of dilation.

Table 4.42 Input parameters used in the models.

Rock type	τ_{sm} (MPa)	ϕ_{0m} (°)	σ_t (MPa)	E (GPa)	ν	Density (kg/m³)	Dilation angle (°)
Coal	1.6	29	0.63	3.33	0.25	1500	0
Sandstone	3.2	34	1.3	10.0	0.25	2500	0

Table 4.43 Change in τ_{sm} and ϕ_{0m} with shear strain.

Shear strain	Cohesion (τ_{sm}) (MPa)	Friction angle (ϕ_{0m}) (°)
0.000	τ_{sm}	ϕ_{0m}
0.005	$\tau_{sm}/5$	$\phi_{0m} - 2.5$
0.010	0	$\phi_{0m} - 5$
0.050	0	$\phi_{0m} - 5$

All these properties were to be specified for the coal seams and the respective roof and floor rocks. The input parameters used for the strain-softening models are as given in Table 4.42.

The residual values for the cohesion and friction angles were taken as:

$$\tau_{sm \text{ (residual)}} = 0 \tag{4.15}$$

$$\phi_{0m \text{ (residual)}} = \phi_{0m} - 5° \tag{4.16}$$

The variations of the cohesion and friction angles with respect to shear strain are given in Table 4.43.

The *in situ* stress field applied in the models was

$$S_H = S_h = 2.4 + 0.01H \text{ MPa} \tag{4.17}$$

$$S_v = 0.025H \text{ MPa} \tag{4.18}$$

where S_H and S_h are the major and minor horizontal *in situ* stress, and S_v is the vertical stress.

An estimation of pillar strength had been made in a way analogous to that of laboratory estimation of uniaxial compressive strength under servo-controlled testing conditions. However, apart from differences in size between real pillars and laboratory specimens so tested, the latter disregarded *in situ* stresses as well as roof and floor behaviour. The method of modelling consisted of the different steps:

1 grid generation,
2 selection of appropriate model,
3 incorporation of material properties, *in situ* stresses and boundary conditions,
4 development of roadway excavations in the model,

5   application of a constant vertical velocity at the top of the model and continuous monitoring of the average vertical stress and strain in the pillar at each solution 'step'.

To define the ultimate load-bearing capacity or the peak stress and to obtain the post-failure behaviour of the pillar, a strain-softening material model had been chosen. The relevant material properties and *in situ* stresses employed had already been discussed. After making the roadway excavations, the top of the model was fixed in the vertical direction to maintain a constant vertical velocity. The other boundary conditions included zero vertical displacements at the model bottom and zero normal displacements at the four vertical symmetry planes. The top of the model was fixed in the vertical direction and a constant velocity of 10^{-5} m/s was applied. The average vertical stresses in all pillar zones and the average vertical strain in the pillar were continuously monitored and plotted. The vertical strain in the pillar was calculated as the average roof-to-floor movement (convergence) over the pillar area divided by the nominal pillar height. It may be noted that peak value of the strain-stress relation curve defines the pillar strength.

4.4.5.1   *Results*

Figure 4.53(a) shows the stress-strain curves obtained for 5 m long pillar and Figure 4.53(b) shows that of an equivalent 10 m × 10 m square pillar when the extraction height was 2.0 m, with a slenderness ratio of 2.5 m for a long pillar. The peak stress values had been found to be in good agreement with each other, confirming the correctness of the equivalent width concept at the W/h ratio.

For a working height of 4 m, with a W/h ratio of 1.25 (in long pillars), Figures 4.54(a) and (b) show similar stress-strain curves with comparable peak stress values.

However, for a 6 m working height, with a slenderness ratio of 0.83, the long pillar of 5 m width had a lower strength compared to that of the equivalent width 10 m × 10 m square pillar as seen in Figures 4.55(a) and (b). A comparison of the peak stress values showed the long pillar failed at about 11% lower stress value compared to the equivalent square pillar.

When the working height was further increased to 8 m, amounting to a slenderness ratio of 0.625, as seen from Figures 4.56(a) and (b), the long pillar had a further reduced strength value as compared to its equivalent width square pillar. The reduction in strength for this W/h ratio was about 15%.

The variation of the peak stress value with respect to the slenderness ratios for long pillars and equivalent square pillars is given in Figure 4.57.

From the above numerical modelling studies the following adaptations were made to the empirical pillar design to account for the effect of the slenderness ratio on the concept of equivalent width:

- For long pillars of width W, having W/h ratios greater than 1.0, the equivalent width (W_e) vis-à-vis the width of a square pillar was taken as $2W$.
- When the W/h ratio was less than or close to 1.0, to err on the side of caution, the strength estimated using Equation (4.2) using $W_e = 2W$ needed to be reduced by 20%.

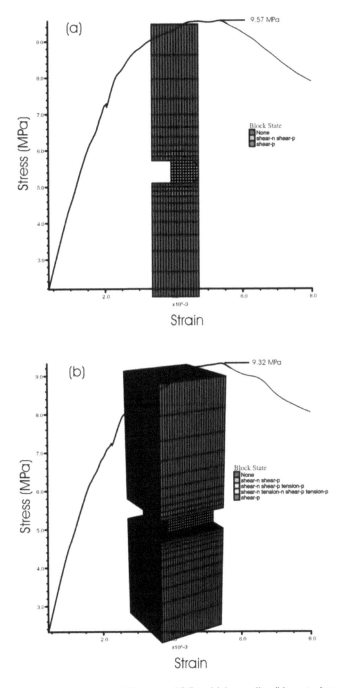

Figure 4.53 Stress-strain curves for W/h ratio of 2.5 in (a) long pillar (b) equivalent square pillar.

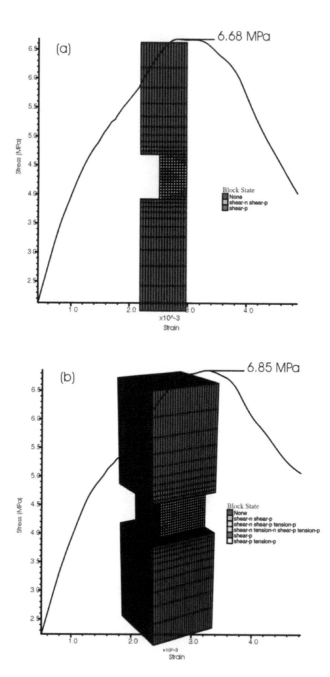

Figure 4.54 Stress-strain curves for W/h ratio of 1.25 in (a) long pillar (b) equivalent square pillar.

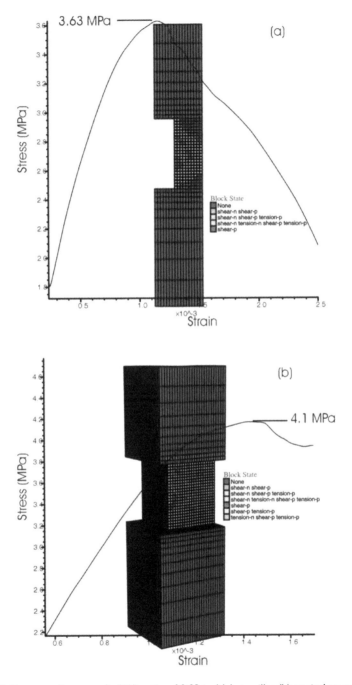

Figure 4.55 Stress-strain curves for W/h ratio of 0.83 in (a) long pillar (b) equivalent square pillar.

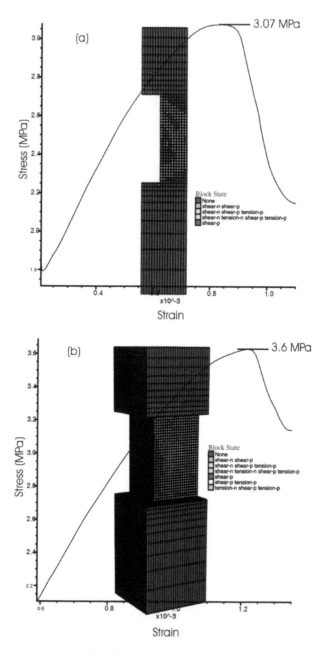

Figure 4.56 Stress-strain curves for W/h ratio of 0.625 in (a) long pillar (b) equivalent square pillar.

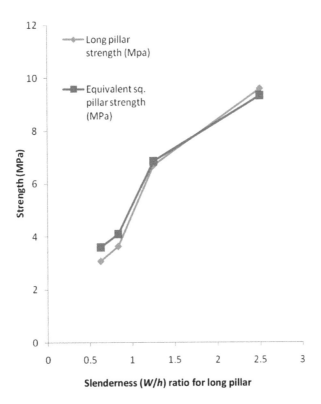

Figure 4.57 Comparison of long pillar and equivalent square pillar strength for various slenderness ratios (Porathur et al., 2013).

4.4.6 Web pillar design for Quarry SEB

As mentioned earlier, the mining areas in Quarry SEB were demarcated into panel A, panel B and panel C as shown in Figure 4.51. The depth regimes at which the targeted seams occurred in all the panels were more or less the same.

The web pillars in all the panels were designed for medium-term stability, i.e. for a few years, so as to stand until the upper seams would have been extracted. Deviation in entries amounting to 0.2% of their length on both sides had also been considered. Therefore, the minimum and maximum web pillar widths would be $(W - 0.4 \times L/100)$ and $(W + 0.4 \times L/100)$, respectively. The web pillar width W had been designed so as to have a minimum safety factor of 1.5. A summary of the required web pillar thicknesses for each workable seam for panels A, B and C is given in Table 4.44.

4.4.7 Web pillar design in Quarry AB below Banji village

The entire area was taken as one panel and the web pillars were designed in individual seams with the minimum required safety factor as 2.0. A summary of the web pillar widths for the proposed workings below the Banji village is given in Table 4.45.

Table 4.44 Design table for panel A, panel B and panel C at Quarry SEB.

Seam	Extraction height (m)	Max. length of extraction (m)	Max. width deviation (±0.4%) (m)	Web pillar width (m)	Min. & max. safety factors	% extraction	Remarks
IX	3.2	190	0.76	2	2.27 & 3.02	64	Only where a minimum 15 m hard cover exists, strength reduced by 20%
X U	1.4	245	0.98	4.3	1.50 & 1.98	45	XU and XL columnised;
X L	1.2	245	0.98	4.3		45	4 m effective extraction height, descending order
IX	2.0	255	1.02	5.5	2.90 & 3.78	39	IX and VIII columnised,
VIII	4.2	255	1.02	5.5	1.50 & 1.89	39	descending order
VII	8.0	275	1.1	11	1.50 & 1.74	24	
VI	4.0	295	1.2	7.2	1.50 & 1.93	33	
V	4.0	310	1.25	7.8	1.50 & 1.89	31	

Table 4.45 Web pillar widths for Banji village at Quarry AB.

Seam	Extraction height (m)	Max. length of extraction (m)	Max. width deviation (±0.4%) (m)	Web pillar width	Min. and max. safety factor	% extraction	Remarks
VIII	3.0	500	2	3	2.69 & 5.05	54	Only where minimum 15 m hard cover exists, strength reduced by 20%
VII	8.0	500	2	7	2.24 & 3.12	33	Strength reduced by 20%
VI	4.5	500	2	6.2	2.00 & 3.04	36	
V	3	500	2	5.6	2.00 & 3.25	38	

4.4.8 Numerical modelling for analysis of overall stability

The overall stability, including the roof stability of entries, parting stability wherever applicable and pillar stability, had been analysed using elastic numerical modelling studies. Two models were constructed, one representing panel A, B and C of Quarry SEB and the other for Banji village in Quarry AB. These models had geo-mining parameters taken from the above design Tables 4.44 and 4.45, for Quarry SEB and Quarry AB, respectively.

Numerical modelling was conducted using *FLAC3D*. The safety factor block contours presented were evaluated using a programming environment (*FISH*) of the above software.

4.4.8.1 Rock properties used for modelling

Physico-mechanical properties used in the elastic models are those provided by M/s Tata Steel Ltd., as given in Table 4.41. The RMR values taken for coal and parting rock were the average values.

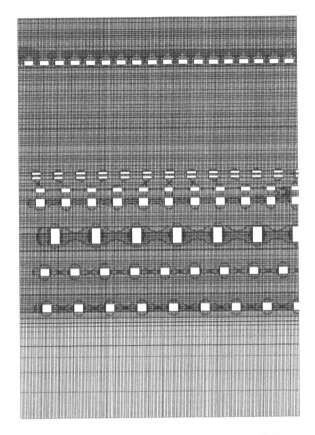

Block contours of
safety factor

☐ 0.0 to 0.5
☐ 0.5 to 1.0
☐ 1.0 to 1.5
☐ 1.5 to 2.0
☐ 2.0 to 2.5
☐ 2.5 to 3.0
☐ 3.0 to 3.5
☐ > 3.0

Figure 4.58 Block contours of safety factor in the domain modelled for Quarry SEB.

4.4.8.2 Modelling procedure

Numerical modelling studies had been envisaged in the following stages:

Stage 1: The 2D models were constructed considering a plane-strain scenario. The virgin stage was modelled, loaded with the elastic properties of the different rock types and *in situ* stresses as given for strain-softening modelling.

Stage 2: Roadways in each seam were excavated and the model was run to equilibrium.

Stage 3: The *FISH* program was run using the output of the numerical model along with strength parameters of different rock types to analyse the roadway, parting and pillar stability in terms of safety factor contours.

4.4.9 Results

Figure 4.58 shows the block contours of the safety factor through the domain modelled for Quarry SEB. A closer view of the safety factor contours in seam XI is given in Figure 4.59(A), in seams XU, XL, IX and VIII is given in Figure 4.59(B), in seam VII in Figure 4.59a(C), and in seams VI and V in Figure 4.59a(D). The roofs of all the web cuts were found to be stable from a short-term point of view. The roof stability was

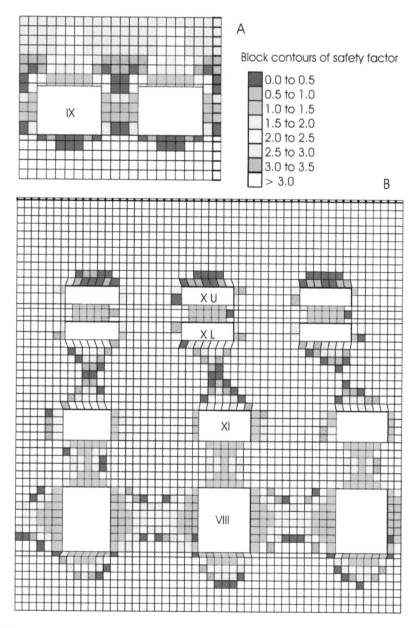

Figure 4.59 Closer view of safety factor contours in (A) seam XI and (B) seams XU, XL, IX and VIII of Quarry SEB.

needed only for a couple of days (short-term) when the CHM would be operating in a web cut.

From Figure 4.59(B) it may be seen that the parting between XL and XU also had good stability, however due to the inconsistency in its thickness, it would be safer to design the pillars in seam X assuming an eventual collapse of the thin parting.

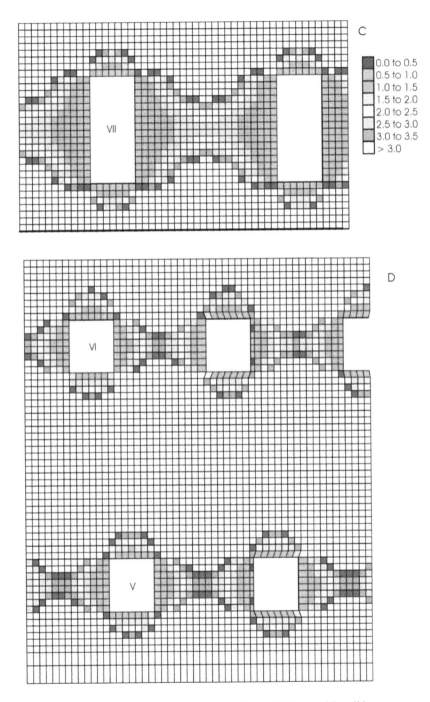

Figure 4.59a Continued. (C) seam VII and (D) seams VI and V.

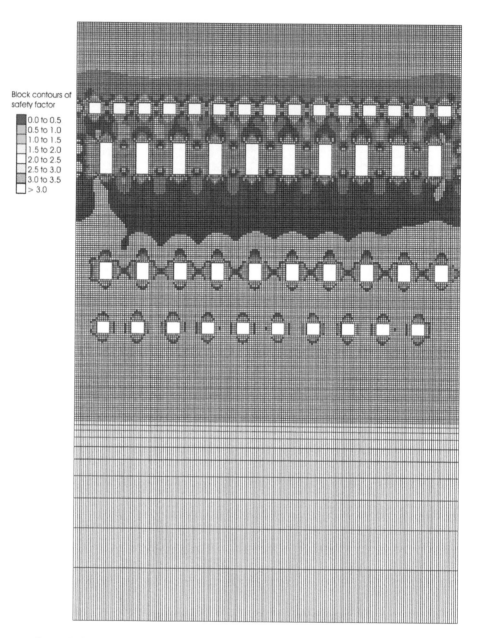

Block contours of
safety factor

- 0.0 to 0.5
- 0.5 to 1.0
- 1.0 to 1.5
- 1.5 to 2.0
- 2.0 to 2.5
- 2.5 to 3.0
- 3.0 to 3.5
- > 3.0

Figure 4.60 Block contours of safety factor in the domain modelled for Banji village.

A minimum parting of 6–7 m between seam XL and seam IX had been modelled as shown in Figure 4.59(B). The safety factor contours in this parting revealed good stability and therefore columnisation of pillars was not found necessary.

Figure 4.60 shows the overall stability of the entries, parting and the pillars in all the proposed seams below Banji village. A closer view of the workings as given in Figures 4.61(A) and (B) shows quite good roof stability of the roadways in all the

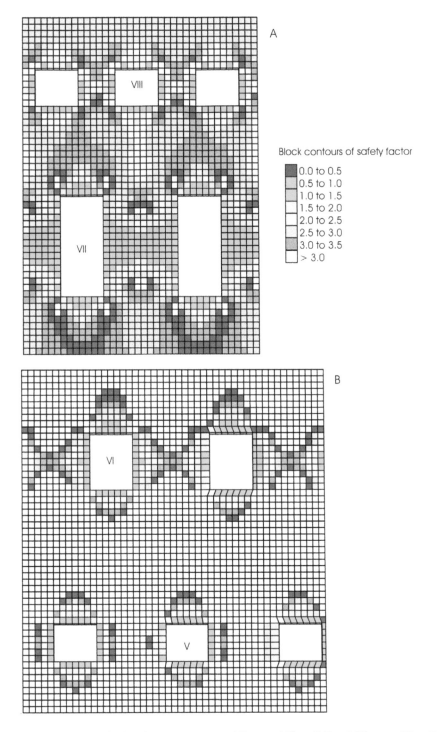

Figure 4.61   Closer view of safety factor contours in (A) seams VIII and VII and (B) seams VI and V of Quarry AB below Banji village.

Figure 4.62 View of highwall mining operation in Quarry SE, West Bokaro of Tata Steel Ltd.

seams with local safety factors in all roof elements greater than 1.0. Figure 4.62 gives an overview of the highwall mining operation in Quarry SE, West Bokaro of Tata Steel Ltd. while Figure 4.63 shows the cutter of CHM capable of 3.5 m wide being used for highwall entries by M/s Cuprum Bagrodia Ltd.

4.4.10   Stability of the final pit slope

As stated in the introduction, CSIR-CIMFR had studied the slope stability at Quarry AB as well as Quarry SEB and the recommendations given in the reports CSIR-CIMFR, 2000 for Quarry AB and CSIR-CIMFR, 2008b for Quarry SEB had to be followed during the extraction of the seams by highwall mining.

4.4.11   Entry mouth stability

From past experiences in the USA and Australia, several cases of web pillar collapses at the entry mouth had occurred (Zipf and Bhatt, 2004). The primary reasons might be the very steep highwall angles and insufficient web thickness at the entry mouth. When the highwall slope is very steep (70–90°), which is the case in most of their mines, it tends to throw a higher cover pressure right at the beginning of the entry. The web pillars have three free faces at the entry mouth, which makes the webs vulnerable to spalling and crushing at the mouth.

At West Bokaro, the recommended highwall slope angle at both the pits was much gentler at about 47°. At the entry mouth region, the cover pressure would therefore be very nominal and therefore the entries were expected to be very stable.

Figure 4.63 View of cutter in CHM capable of 3.5 m wide web cuts (Courtesy: Cuprum Bagrodia Ltd.).

4.4.12 Findings and conclusions

- From the input parameters collected from the mine site and those supplied by TSL, the seams feasible for highwall mining extraction at Quarry SEB included seams XI, XU, XL, IX, VIII, VII, VI and V.
- Seams VIII, VII, VI and V were workable by highwall mining at Quarry AB below Banji village.
- The proposed highwall mining area in Quarry SEB was divided into three panels: panel A, B and C, as shown in Figure 4.51. However, due to very little depth and thickness variation throughout the quarry, a common design was used for Quarry SEB.
- However, the top seams in both the quarries namely seam XI in Quarry SEB and seam VIII below Banji village, could be extracted only after a more accurate assessment of the existence of a minimum 15 m hard cover in those regions.
- The sequence of extraction of seams at all sites was from bottom to top.
- The thickness of the web pillars to be left in each seam at Quarry SEB and Quarry AB below Banji village is given in Tables 4.44 and 4.45 respectively.
- The web pillars at Quarry SEB were designed for a minimum safety factor of 1.5 due to the absence of any permanent surface features to be protected from subsidence.

- Due to the densely populated Banji village area, the web pillars in Quarry AB were designed for a minimum safety factor of 2.0 for the protection of surface structures on a long-term basis.
- Elasto-plastic numerical modelling studies indicated that when the slenderness ratio (W/h ratio) of a long pillar goes below 1.0, its strength might be reduced up to 15% as compared to its equivalent square pillar of width $2W$. For the top seams in Quarry SEB and below Banji village, wherever the slenderness ratio was less than or closer to 1.0, the strength had been reduced by 20%.
- There was no requirement of barriers or skip holes within the highwall mining area, since the minimum safety factor considered in all places was 1.5.
- At Quarry SEB, seams XU and XL were to be extracted simultaneously with columnised pillars. The upper section needed to be extracted first followed by the corresponding lower section. The thin parting between the two seams was found to be unstable and also had inconsistent thickness, so the pillars in these two seams were designed anticipating the non-existence of the parting and considering a total working height of 4.0 m.
- Similarly, the seams IX and VIII also needed to be extracted simultaneously with vertical superposition of the workings at Quarry SEB. Here also, the upper seam IX needed to be extracted first followed by the lower seam VIII. However, the parting between the two seams was anticipated to be stable.
- Wherever previously worked underground roadways exist, highwall mining in the same horizon should be done after leaving a safety barrier of width not less than that of the web pillar. This was to make sure that the highwall web cuts would not run into the previous workings.
- For stabilisation of the final pit slopes, the recommendations given in the earlier reports of CSIR-CIMFR (CSIR-CIMFR, 2000 for Quarry AB and CSIR-CIMFR, 2008b for Quarry SEB) were followed.
- Elastic numerical modelling and safety factor contouring studies indicated satisfactory roof stability for all the workings on a short-term basis, for the time the machine would be operating for a single cut.

4.4.13   Impact of surrounding blasting on highwall mining

Blasting activities close to the proposed highwall mining at Quarry South-East Barrier (Quarry SEB) and Quarry AB could affect the stability of the highwall mining entries, web pillars as well as the highwall slopes. Ground vibration generated from the day-to-day blasting operation nearby the highwall mining entries could cause premature roof and side collapse, thereby, trapping the continuous miner inside the highwall mining entries. Flyrock generated from the blasting nearby the highwall mining could also damage the equipment used for the mining such as the launch vehicle etc. and cause safety concerns for the workers.

4.4.13.1   Field study and ground vibration results

The data of the field investigations carried out in Quarry SEB, West Bokaro of TSL in 2004 to assess the impact of blast-induced ground vibrations, air overpressure/noise, flyrock on the nearby village structures within 100 m from the blasting source (CSIR-CIMFR, 2004) were used for the purpose of determining the blasting impact on highwall mining. This was the same site where highwall mining was proposed. The research team monitored ground vibrations and their dominant frequencies, air over-pressures/noise generated due to eleven production blasts at different working benches of the mine. The total explosive charge fired in a round of blast varied between 2.278 and 10.820 tonnes. The maximum charge per delay varied between 120.00 and 508.00 kg. The blasthole diameter used was 160 mm. Hole depths varied between 10.0 and 14.8 m for XI OB-1 bench and for XI OB-2 it varied between 9.0 and 17.4 m. For drainage blasting, the depth of the holes varied between 5.8 and 8.2 m. The Nonel system of IEL, Orica, was used for in-hole as well as surface initiation in all the blasts at the mine. The distances of vibration monitoring points for the different blasts varied between 75 and 250 m.

Thirty-three ground vibration data were collected during the one year period of field investigation. The magnitude of ground vibration recorded varied between 5.36 and 54.08 mm/s. The maximum magnitude of vibrations recorded during the period of the field investigation was 54.08 mm/s at a distance of 75 m from the blast with an associated dominant peak frequency of 9.6 Hz. The maximum explosives weight per delay was 455 kg and the total explosive fired in the round was 9,318 kg. The frequency levels recorded from the different blasts varied widely, ranging from 8.0 to 39 Hz. FFT analyses of the vibration data revealed that in the majority of the blasts, the concentration of vibration energy ranged between 10–24 Hz. Air overpressure values recorded during the period of investigation varied from 118.0–134.0 dB (L). No flyrock was observed or recorded in any of the experimental blasts conducted.

4.4.13.2   Assessment of safe value of PPV

As per DGMS, 2007 (Tables 4.18 and 4.19), the permissible limit of peak particle velocity for the stability of the roof and web pillars of the proposed highwall mining depends on the RMR values of the immediate roof. The details of various seams in the highwall mining area are given in Tables 4.46 and 4.47. Seam VII was the main targeted seam. The nature of the roofs and their respective RMR values for different seams where highwall mining was proposed are given in Table 4.48.

These RMR values were provided by TSL. The lowest RMR value from the different coal seams was 57 for seam VI. Seams VII and VIII had similar values of RMR (62 each) and seams V and IX had similar RMR values (71 each). Therefore, considering the different values of RMR for different seams, the threshold values of peak particle velocity for the safety of highwall mining entries had been determined. The safe values of PPV for different roofs of highwall entries are given in Table 4.49.

Based on the generalised criteria proposed by Bauer and Calder (1971) (Table 4.20) for damage level of particle velocity on rock mass and slopes, the maximum value of the threshold PPV (i.e. 120 mm/s) suggested for the safety of the roofs of the highwall entries would also be applicable for highwall slopes (Table 4.18). The highwall slopes would also have less damage impact due to ground vibration in comparison to the roof of the galleries.

Table 4.46 Details of various seams in the Banji village area.

Unit	Thickness (m)	Average thickness (m)	Approximate depth (m)
Seam X	1–3	1.5	14
Seam X/Seam IX IB	0–0.7	0.5	–
Seam IX	0–2	1.0	16
Seam VIII/Seam IX IB	2.6–3.3	2.9	–
Seam VIII	1–4.5	3.0	22
Seam VII/Seam VIII I B	7–13	10.0	–
Seam VII	8	8.0	35
Seam VI/VII IB	14–25	18.0	–
Seam VI	5	5.0	60
Seam V/VI IB	5–18	13.0	–
Seam V	3	3.0	80

Table 4.47 Details of various seams in the Quarry SEB area.

Unit	Thickness (m)	Average thickness (m)	Approximate depth (m)
Seam XU	1–2	1.5	50
Seam XL/Seam XU IB	0–2	1.0	–
Seam XL	0–5	1.2	50
Seam IX/Seam XL I B	7–20	12.0	–
Seam IX	1–2	1.5	64
Seam VIII/Seam IX I B	2–4	3.0	–
Seam VIII	1–6	4.5	69
Seam VII/Seam VIII IB	11–30	20.0	–
Seam VII	6–11	8.0	95
Seam VI/Seam VII IB	10–20	15.0	–
Seam VI	1–5	4.0	120
Seam V/Seam VI IB	5–20	13.0	–
Seam V	2–10	4.5	140

Table 4.48 Nature of roof and corresponding RMR values of different seams.

Name of seam	Nature of the immediate roof	RMR values
V	Sandstone	71
VI	Sandstone	57
VII	Shale	62
VIII	Carbonaceous shale	62
IX	Shale	71
X U	Carbonaceous shale	65

4.4.13.3 Assessment for safe blasting parameters

The ground vibration data recorded during the field investigations were grouped together for statistical analysis. An empirical equation had been established correlating the maximum explosive weight per delay (Q_{max} in kg), distance of vibration

Table 4.49 Safe values of PPV for roofs of highwall entries corresponding to the RMR values of different seams at West Bokaro, TSL.

Name of seam	RMR values	Safe value of PPV (mm/s)
Seams V & IX	71	120
Seam VI	57	114
Seams VII & VIII	62	120
Seam X U	65	120

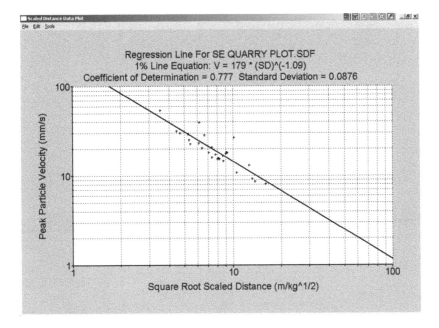

Figure 4.64 Regression plot of vibration data recorded at various locations due to blasting at Quarry SEB, TSL.

measuring transducers from the blast face (D in m) and recorded peak particle velocity (V in mm/s). A regression plot of the vibration data recorded at various locations due to blasting at Quarry SEB, TSL is given in Figure 4.64. The established equation for the site is:

$$V = 179[D/\sqrt{Q_{max}}]^{-1.09} \qquad (4.19)$$

Coefficient of determination $= 0.777$
Standard deviation $= 0.0876$

The above equation could be applicable for Quarry SEB as well as Quarry AB, West Bokaro, TSL. This equation could be used to compute the safe maximum explosive

Table 4.50 Safe explosive charge per delay at various distances from the blasting site for seams V, VII, VIII, IX & X.

Distance from the blasting face (m)	Suggested maximum charge per delay (kg)
50	1,208
75	2,718
100	4,832
125	7,550
150	10,873
175	14,799
200	19,329
225	24,463
250	30,202
275	36,544
300	43,491

Table 4.51 Safe explosive charge per delay at various distances from the blasting site for seam VI.

Distance from the blasting face (m)	Suggested maximum charge per delay (kg)
50	1,100
75	2,474
100	4,399
125	6,874
150	9,898
175	13,472
200	17,596
225	22,270
250	27,494
275	33,268
300	39,592

weight to be detonated in a delay for distances of concern. Bearing in mind the safe peak particle velocity values for the total safety of highwall mining entries, the safe levels of maximum explosive weight per delay had been calculated for various seams and for various distances of concern, using Equation (4.19) and are given in Tables 4.50 and 4.51. The same are also represented in Figures 4.65 and 4.66. However, for improved safety, it was always advisable to contain the total firing time (i.e. difference between the firing times of the last hole to the first hole) in the blasting round within 1000 millisecond i.e. 1 second.

4.4.13.4   Summary of findings and recommendations

The following conclusions were drawn on the basis of the field study at Quarry SEB, TSL, of the stability of the highwall mining entries, web pillars and highwall slopes from blast-induced ground vibrations.

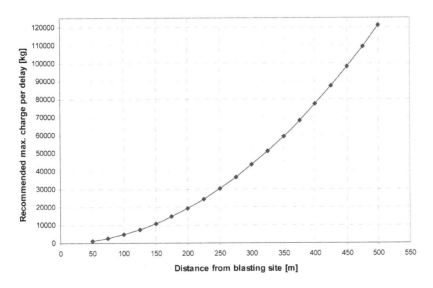

Figure 4.65 Recommended charge per delay to be fired in a blasting round for the safety of highwall mining entries (seams V, VII, VIII, IX and X).

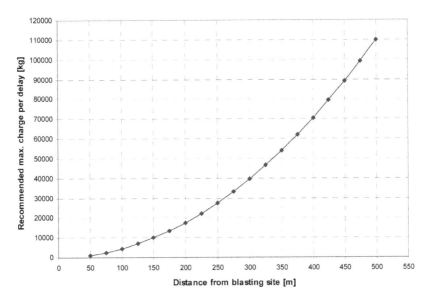

Figure 4.66 Recommended charge per delay to be fired in a blasting round for the safety of highwall mining entries (seam VI).

1   Blasting activities close to the proposed highwall mining at different mining areas of West Bokaro Colliery could affect the stability of highwall mining entries, web pillars and the highwall slope. As such, a controlled blasting operation had to be carried out at such locations.

2   Eleven production blasts were monitored during the field investigation at the Quarry SEB, West Bokaro, with varying designs and charging patterns. The total explosive charge fired in a round of blast varied between 2.278 and 10.820 tonnes. The maximum charge per delay varied between 120.00 and 508.00 kg. The blasthole diameter used was 160 mm.

3   The lowest RMR value from the different coal seams found in the area was 57 (seam VI). Seams VII and VIII had similar values of RMR (62 each) and seams V and IX had similar RMR values (71). Therefore, considering the different values of RMR for different seams, the threshold values of peak particle velocity for the safety of highwall mining entries were determined. The safe values of PPV for different roofs of highwall entries are given in Table 4.49.

4   Based on the generalised criteria proposed by Bauer and Calder (1971) (Table 4.20) for damage level of particle velocity on rock mass and slopes, the maximum value of the threshold PPV (120 mm/s) suggested for the safety of the roof of the highwall entries would also be in safer side for highwall slopes. The highwall slopes also had less damage impact to ground vibrations in comparison to the roof of the galleries.

5   The propagation equation for prediction of blast-induced ground vibration at SE Quarry had been established and is given as Equation (4.19). Bearing in mind the safe peak particle velocity values for the total safety of highwall mining entries, the safe levels of maximum explosive weight per delay had been calculated for various seams and for various distances of concern, using Equation (4.19) and are given in Tables 4.50 and 4.51. The same are also represented in Figures 4.65 and 4.66. For improved safety, it was always advisable to contain the total firing time (i.e. difference of the firing times of the last hole to the first hole) of the round within 1000 millisecond i.e. 1 second.

6   More refined predictability necessitates further experiments at Quarry SEB and AB which would improve the predictive capability of Equation (4.19).

4.5   SITE D

TRENCH-CUT HIGHWALL MINING AT SHARDA OPENCAST PROJECT OF M/S SOUTH EASTERN COALFIELDS LIMITED (SECL)

At the Sharda Highwall Mining Project of Sohagpur area, SECL, thin coal seams were extracted using a trench highwall mining method. It was important to optimise the trench dimensions to minimise the land degradation and volume of trench excavation and at the same time maintain safety during the entire highwall mining operation.

A parametric study of slope stability had been done for finding the effects of slope angle and groundwater conditions on trench slope stability. An instrumentation scheme was proposed to monitor the slope surface, web pillars, overburden strata movement and surface subsidence during and after the highwall mining extractions. In a previous scientific study of CSIR-CIMFR (CSIR-CIMFR, 2014) the trench stability for a given slope angle of 70° was analysed and the stability was found to be satisfactory.

4.5.1 Scope of work

To visualise the stability of the trench slope for different slope angles, six slope angles 45°, 60°, 65°, 70°, 80° and 90° were considered and simulated considering each excavations stage by stage from the excavation of the top seam (seam VI Bottom) to the bottom seam (seam IV). The final pit bottom of the slope for the six slope angles was considered for a slope stability assessment in order to evaluate the final pit stability.

To analyse the effects of groundwater on trench slope stability, a slope stability assessment in three dimensions had been done which considered the drained rock mass condition as well as the undrained/saturated rock mass conditions.

An instrumentation scheme was suggested in order to monitor the slope surface and trench for surface movements. This consisted of monitoring displacements within the overburden strata and slope surface as well as the load on web pillars.

4.5.2 Numerical modelling studies

4.5.2.1 Methodology

Numerical modelling in three dimensions has been done to analyse the effects of slope angle and effects of groundwater on the overall stability of the trench slope during the final stages of extractions i.e. extraction of seam IV when the proposed trench has the highest depth.

Initially the pit slope was analysed with an angle of 70° in the hard rock portion and an angle of 37–45° in the overlying settled OB dump. The final trench slope for the above slope angle was found to have a safety factor (or Factor of Safety) of 1.52 after all extractions were done in seam IV (CSIR-CIMFR, 2014).

Three-dimensional numerical modelling in *FLAC3D* had been done considering six slope angles 45°, 60°, 65°, 70°, 80° and 90° for the final trench slope in order to visualise the effects of the slope angle.

These models considered saturated rock mass conditions (which was in a way the worst-case scenario). Further, groundwater conditions had been analysed considering the pore water pressure with a dry condition of rock mass and incorporating a water table with different densities below (saturated) and above (unsaturated) the water levels. These models for studying the effects of groundwater were performed only for 70° slope angle configuration.

The rock mass properties taken into consideration for trench stability assessment in drained and saturated condition are as given in Table 4.52.

Table 4.52 Physico-mechanical properties considered for modelling.

Physico-mechanical properties								
Material	Young's modulus (Pa)	Poisson's ratio	Density (kN/m³)		Cohesion (kPa)		Friction (°)	
			Dry	Saturated	Saturated	Dry	Saturated	Dry
Soil	2.50E+08	0.1	1.90E+01	1.90E+01	–	50	–	25°
Sandstone (parting)	1.00E+09	0.2	2.40E+01	2.50E+01	250	500	35°	40°
Coal	5.00E+08	0.2	1.50E+01	1.60E+01	250	500	35°	40°

Figure 4.67 Final trench slope geometry for slope angle 45°.

4.5.2.2 *Effect of slope angle*

The effect of the slope angles had been analysed by comparing the safety factor values for three different slope angles. When the slope angle of the final trench slope was 45°, the pit geometry would be as shown in Figure 4.67. The failure surface of the slope after all excavations had been done was as given in Figure 4.68. The safety factor of the trench slope for 45° was 2.0 as given in Figure 4.68. However, this safety factor was indicative of the weakest portion which was the top settled OB dump. This means the overall slope safety factor passing through the hard rock mass was definitely > 2.0.

When the slope angle of the pit was 60°, the safety factor of the slope was 1.75 and the failure surface of the pit was as shown in Figure 4.69. The weakest zone passes through the slope crest to toe on the settled OB dump were as shown in Figure 4.69.

When the slope angle of the pit was 65°, the pit geometry was as given in Figure 4.70. The failure surface could be seen going through the slope crest to toe with a safety factor of 1.61 as shown in Figure 4.71.

When the slope angle of the pit was 70°, the geometry of the pit was as shown in Figure 4.72. The failure surface could be seen going through the slope crest to toe as shown in Figure 4.73. The safety factor of the final trench was 1.52 as given in Figure 4.73.

When the slope angle of the pit is 80° the geometry of the pit is as shown in Figure 4.74. The failure surface can be seen going through the hard rock portion and reaching the slope toe as shown in Figure 4.75. The safety factor of the final trench is 1.35 as given in Figure 4.75.

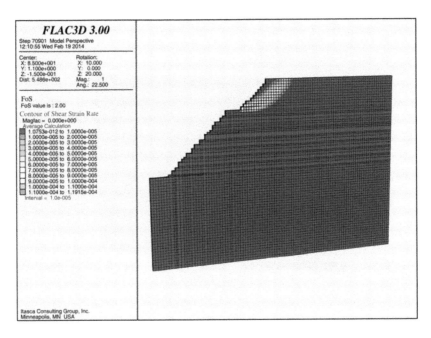

Figure 4.68 Failure surface and safety factor (i.e. FoS) for slope angle 45°.

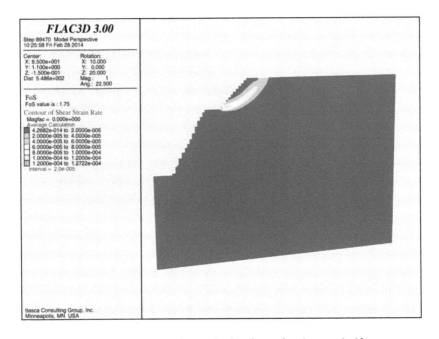

Figure 4.69 Failure surface and safety factor for slope angle 60°.

Figure 4.70 Final trench slope geometry for slope angle 65°.

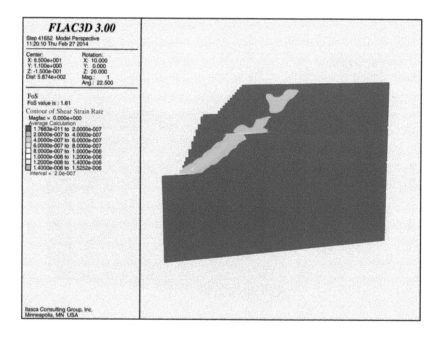

Figure 4.71 Failure surface and safety factor for slope angle 65°.

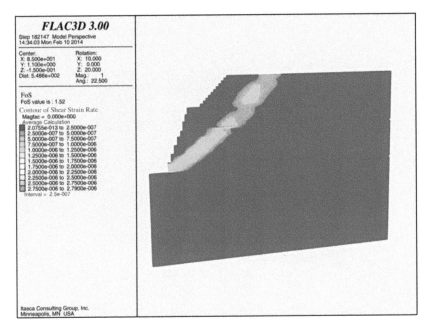

Figure 4.72 Final trench slope geometry for slope angle 70°.

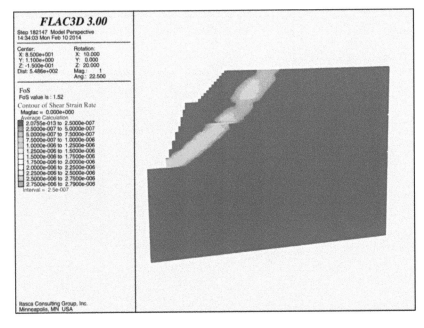

Figure 4.73 Safety factor and failure surface for slope angle 70°.

Figure 4.74 Final trench slope geometry for slope angle 80°.

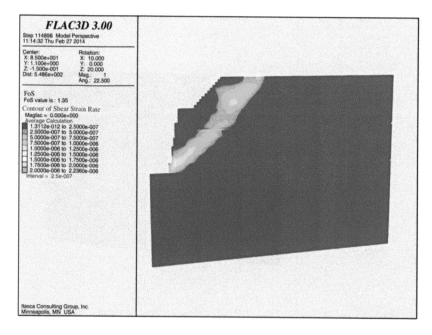

Figure 4.75 Safety factor and failure surface for slope angle 80°.

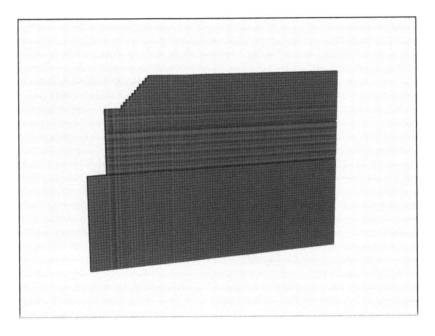

Figure 4.76 Final trench slope geometry for slope angle 90°.

When the slope angle of the pit was 90°, the geometry of the pit was as shown in Figure 4.76. The failure surface could be seen going through the hard rock portion and reaching the slope toe as shown in Figure 4.77. The safety factor of the final trench was 1.10 as given in Figure 4.77. It could also be concluded that the safety factor dropped considerably as the trench slope was vertical, with the steep face of rock mass having a tendency to slide or topple.

The variation in the safety factor of the trench slope for the four slope angles 45°, 60°, 70° and 90° is plotted and given in Figure 4.78. The variation in safety factor clearly indicated that any angle <80° might be acceptable from a slope stability point of view, as the required safety factor of a trench less than 2–3 years old could be considered as 1.3. Therefore a trench slope of 70° might be more suitable to be on the safer side.

4.5.2.3 *Effects of rock mass condition*

To analyse the effects of the rock mass condition, the properties pertaining to saturated and drained conditions were used for simulation. The recommended final trench slope with a slope angle of 70° was simulated using the drained rock mass strength properties as given in Table 4.50.

Numerical modelling was done with stage by stage excavations from the opening of the trench up to the final stage extractions at seam IV. The failure surface and safety factor for drained rock mass strength properties are shown in Figure 4.79.

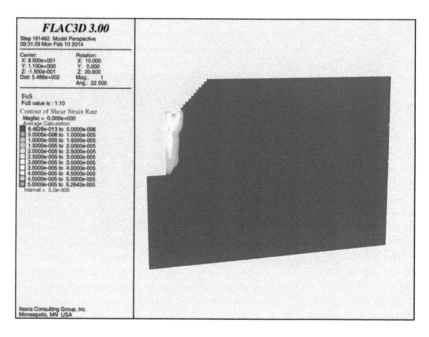

Figure 4.77 Safety factor and failure surface for slope angle 90°.

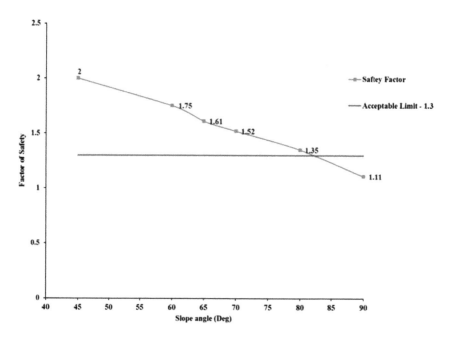

Figure 4.78 Variation in safety factor for different slope angles 45°–90°.

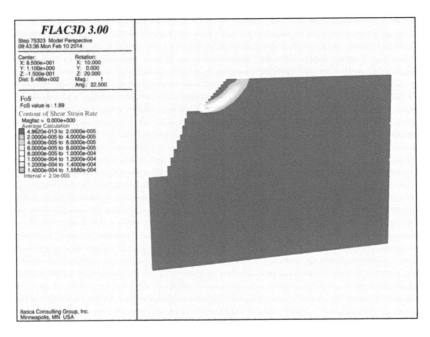

Figure 4.79 Failure surface and safety factor for final trench slope 70° with drained rock mass properties.

Here also the numerical modelling results showed that the rock mass (sandstone) had a much superior stability, with the weakest portion being the settled OB dump region, which had a safety factor of 1.89. This meant that the rock mass in a drained state had a safety factor >1.89 in comparison to the saturated state safety factor of 1.52.

4.5.2.4   *Instrumentation plan*

The instrumentation plan suggested that in order to monitor the stability of the slope as well as the web pillars, the stress cells had to be installed into some of the web pillars to monitor the load on the web pillar. Multi-point borehole extensometers were also recommended to be installed from the surface until the roof of seam IV to monitor the strata movement and surface movement. Total station monitoring (reflectors) were proposed to be installed to monitor the slope movement during the highwall mining operations.

4.5.2.5   *Stress cells in web pillars*

The stress cells were to be installed in the web pillars through boreholes driven parallel to the axis of the web pillars as shown in Figure 4.80 for seam VI Bottom and seam IV web pillars. The length of the boreholes might be up to 10–15 m and the direction of stress measurement would be vertical. At least 5 or 6 stress cells were to be installed into any of the web pillars, prior to the extraction of the adjacent web cuts.

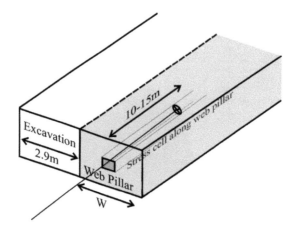

Figure 4.80 Suggested stress cell installation scheme for web pillars at seam VI Bottom and seam IV.

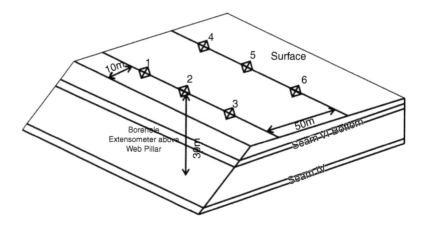

Figure 4.81 Suggested scheme of borehole extensometers to be installed from the surface.

The stress cells should be of vibrating-wire type, for soft rock, with a hole diameter of 38–42 mm, and should have a capacity of measuring >2 MPa stress change. The frequency of instrument reading should be at least one reading in a week.

4.5.2.6 Borehole extensometers

Multi-point borehole extensometers were suggested as given in Figure 4.81. Up to five anchors per borehole extensometer could be installed to record the strata and surface movement. The mechanical type borehole extensometers were most suitable in that case with a maximum length of 30–50 m, and with anchors spaced at 5–10 m intervals. At least five or six anchors per extensometer were recommended to be placed at various depths from the surface. One anchor should be placed at the roof of the extracted seams and the others might be placed at 5 or 10 m within the overburden strata. The horizon of displacement measurement could be down to the depth of seam

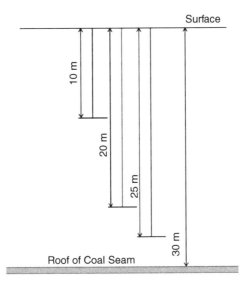

Figure 4.82 The suggested pattern of anchor positions of borehole extensometers.

IV's roof. The total number of extensometers might be 5–6, and they can be placed at a random interval as suggested in Figure 4.81. The pattern of anchor locations of multi-point borehole extensometers should be as given in Figure 4.82. The frequency of instrument reading should be at least one reading in a week.

4.5.2.7 Surface subsidence

It was recommended that the surface subsidence be monitored using total stations with subsidence survey points (survey pillars) installed at the surface along specific survey lines over the extraction panels. The survey stations should be placed at an interval of 10 m from one another.

4.5.2.8 Slope monitoring

Slope monitoring should be done using a total station. Slope monitoring points should be at three locations along a cross section: the crest of the settled OB dump slope, the crest of the rock slope and the horizon of seam VI (bottom) as shown in Figure 4.83. The slope monitoring might be done on a daily basis.

4.5.2.9 Findings and recommendations

From the parametric study of the trench slope stability analysis conducted using three-dimensional numerical modelling for the effects of slope angle and groundwater, the following conclusions have been drawn:

- The overall slope safety factor in a saturated condition for slope angles 45°–90°, showed a variation in safety factor ranging from 1.1 (for 90° trench slope) to >2.0

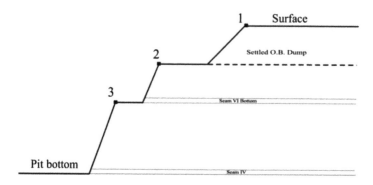

Figure 4.83 Schematic of suggested slope monitoring points up to the top seam horizon.

(for 45° trench slope). A 70° trench slope angle would have a safety factor of 1.52, which was within the acceptable range of 1.2–1.3. A steep slope angle (vertical wall) might result in a very low safety factor, with a tendency to slide or topple.

- A drained rock mass condition would result in a much higher safety factor for the slope. Numerical modelling revealed SF > 1.9 for the rock mass slope, with the weakest region being the settled OB dump. The saturated case (the worst-case scenario) would have a SF of 1.52 for a 70° slope angle.
- Therefore a 70° slope angle was recommended to minimise the land degradation and volume of excavation and also to ensure safety during the highwall mining operations.
- It was recommended that a safety beam of 6.0 m width should be left at the horizon of seam VI's (bottom) floor while deepening the trench to approach seam IV to hold any loose rock falling into the pit. Benches should also be made in the sandstone rock mass, forming an overall slope angle of 70° in the hard rock mass.
- It was recommended that the trench blasting should be conducted with pre-splitting techniques to ensure a smooth surface of the trench walls and to minimise the formation of loose boulders.
- It was also recommended that ground monitoring should be done during highwall mining operations. Four monitoring schemes were proposed: (1) stress cells to be installed into some web pillars for monitoring the vertical stress, (2) multi-point borehole extensometers to be installed from the surface down to the seam horizon to study the strata movement, (3) surface subsidence monitoring and (4) slope monitoring using total station.

4.5.3   Blast design for pre-split blasting

M/s Cuprum Bagrodia Limited awarded the scientific study to CSIR-CIMFR for improvement of blasting efficiency and pre-split blast design for the formation of high-wall benches at Sharda opencast project, South Eastern Coalfields Limited (SECL) (CSIR-CIMFR, 2015). After studying the nature and type of rock strata present at Trench No. 3 (T-3), a smooth blasting technique was implemented for the improvement

Figure 4.84 Joint planes parallel to the highwall slope in top section of 0 to −6 m bench in T-3.

of the powder factor. Rock samples were also collected from a −6 m to −12 m bench of T-3 area for determination of physico-mechanical properties. Based on the nature of rock deposits and their physico-mechanical properties, blast design patterns for pre-split blasting were evolved and tried in in-field conditions for a stable and smooth final wall.

4.5.3.1 Type and nature of rock deposits

The type of rocks present at the T-3 area mainly consisted of massive sandstone. They were classified as:

- Very Coarse-grained sandstone (VCgsst)
- Coarse-grained sandstone (Cgsst)
- Medium-grained sandstone (Mgsst)
- InterCalation of Shale and sandstone (I/C-Shale-sst)
- Shale and coal

During the field visit, only hard rock portions of 0 to −12 m were exposed in the T-3 area. The first bench (i.e. 0 to −6 m) consisted of shaley sandstone and a massive formation of sandstone. The shaley sandstone was present in the top portion of the bench and the thickness varied from 1.5 to 2.4 m. Apart from the lamination (bedding plane), two major joint sets were observed in the top i.e. shaley sandstone portion. One joint plane was found nearly perpendicular to the highwall slope and the other joint set was also nearly parallel to the highwall slope (Figures 4.84 and 4.85). The lower portion of about 3.6 to 4.5 m consisted of white coloured, coarse- to medium-grained sandstone. No prominent joint planes were observed except for the lamination plane.

The second bench (i.e. −6 to −12 m) consisted of hard and massive coarse- to medium-grained white coloured sandstone (Figure 4.86). No prominent joints were visible in the exposed highwall. Rebound hardness values measured on the massive

Figure 4.85 Joint planes perpendicular to the highwall slope in top section of 0 to −6 m bench in T-3.

Figure 4.86 Massive sandstone formation in 0 to −12 m in T-3.

sandstone varied from 22 to 32. In order to understand the nature of the rock that could be encountered in the T-3 area, the exposed rock strata in the T-1 area were observed and studied. It was observed that hard and massive formations of white coloured, coarse- to medium-grained sandstone constituted the major portion of the strata (Figure 4.87). However, shale and shaley sandstone were present on top and bottom of the coal seams which could create problems during pre-split blasting.

4.5.3.2 Physico-mechanical properties of rocks in T-3 area

The rock mass properties of the T-3 area were determined at the CSIR-CIMFR Rock Testing Laboratory. The minimum, maximum and the average values of the uniaxial

Figure 4.87 View of different rock strata along the highwall in T-1.

compressive strength, tensile strength, Young's modulus, Poisson's ratio, shear modulus, bulk modulus, apparent cohesion and angle of internal friction for different rocks are given in Table 4.53. The primary wave velocity (P-wave) of medium-grained sandstone varied between 2564 and 2615 m/s and the shear wave velocity (S-wave) varied between 1442 and 1453 m/s. The average density of medium-grained sandstone was 2.19 g/cc.

4.5.4 Pre-split blast design consideration

In pre-splitting, a fracture plane is made in the rock mass before firing the production blast by means of a row of blastholes with decoupled explosive charges (Pal Roy, 2005). When two explosive charges are initiated simultaneously in the adjacent boreholes, the collision of shock waves between the boreholes causes cracking in tension and produces a sheared zone between the blastholes as shown in Figure 4.88.

The cracking between boreholes, initially produced by the shock wave of the explosive, is subsequently extended and widened by the expanding gases based on three factors: (1) properties and conditions of the rock, (2) spacing between blastholes, and (3) amount and type of explosive in the holes. The fractured zone between the boreholes may be a single, narrow thick zone of fractured rock. This split or crack in the rock forms a discontinuous zone that (1) minimises or eliminates overbreak from the subsequent primary blast and (2) produces a smooth-finish rock wall.

When rock has numerous joints between blastholes and those joints intersect the blasting face at less than 15° angle, it will be impossible to form a good and smooth face with controlled blasting techniques (Konya and Walter, 1990). In a heavily jointed rock mass, pre-splitting may not yield good results whereas smooth blasting may provide better results.

The blasthole diameter for pre-split blasting mainly depends on the availability of drilling machines at the project sites. The blasthole diameter used by different

Table 4.53 Physico-mechanical properties of different rock types at the Sharda project.

Type of rock	Properties	Minimum value	Maximum value	Average value
Very coarse-grained sandstone	Compressive strength (MPa)	2.02	18.59	9.02
	Tensile strength (MPa)	0.36	2.80	1.19
	Young's modulus (GPa)	0.58	5.10	2.92
	Poisson's ratio	0.11	0.30	0.24
	Shear modulus (GPa)	0.29	2.30	1.18
	Bulk modulus (GPa)	0.19	4.17	1.82
	Apparent cohesion (MPa)	0.82	3.90	2.63
	Angle of internal friction (°)	42.49	57.37	48.72
Coarse-grained sandstone	Compressive strength (MPa)	4.00	29.16	12.92
	Tensile strength (MPa)	0.70	3.37	1.52
	Young's modulus (GPa)	0.65	8.70	3.43
	Poisson's ratio	0.05	0.29	0.19
	Shear modulus (GPa)	0.29	3.51	1.48
	Bulk modulus (GPa)	0.30	5.58	1.97
	Apparent cohesion (MPa)	1.35	6.37	3.52
	Angle of internal friction (°)	38.55	56.00	46.63
Medium-grained sandstone	Compressive strength (MPa)	7.64	35.61	15.77
	Tensile strength (MPa)	0.84	3.99	2.23
	Young's modulus (GPa)	1.18	4.98	3.34
	Poisson's ratio	0.04	0.28	0.15
	Shear modulus (GPa)	0.53	2.39	1.44
	Bulk modulus (GPa)	0.52	3.01	1.63
	Apparent cohesion (MPa)	2.33	5.94	3.94
	Angle of internal friction (°)	30.04	58.14	47.23
Intercalation of shale and sandstone	Compressive strength (MPa)	11.39	44.68	21.78
	Tensile strength (MPa)	1.37	5.83	3.65
	Young's modulus (GPa)	1.33	5.88	3.47
	Poisson's ratio	0.02	0.31	0.15
	Shear modulus (GPa)	0.54	2.67	1.52
	Bulk modulus (GPa)	0.69	3.67	1.78
	Apparent cohesion (MPa)	3.61	6.55	4.58
	Angle of internal friction (°)	39.50	55.82	47.54
Shale	Compressive strength (MPa)	16.02	45.34	24.16
	Tensile strength (MPa)	0.90	5.50	3.53
	Young's modulus (GPa)	2.09	5.14	3.82
	Poisson's ratio	0.06	0.25	0.17
	Shear modulus (GPa)	0.98	2.12	1.63
	Bulk modulus (GPa)	0.81	3.06	2.09
	Apparent cohesion (MPa)	......	.....	.....
	Angle of internal friction (°)	.....	.....	.....
Coal	Compressive strength (MPa)	17.75	43.43	31.01
	Tensile strength (MPa)	0.36	4.58	2.27
	Young's modulus (GPa)	1.52	2.06	1.75
	Poisson's ratio	0.04	0.18	0.12
	Shear modulus (GPa)	0.66	0.87	0.78
	Bulk modulus (GPa)	0.58	1.07	0.78
	Apparent cohesion (MPa)	.....	.....	.....
	Angle of internal friction (°)	.....	.....	.....

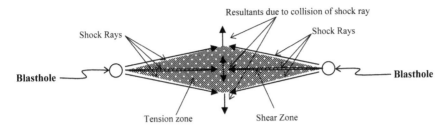

Figure 4.88 Pre-spitting theory illustration (Source: ISEE Blaster's Handbook™, 18th Ed., Figure 36.7).

persons for pre-split blasting varied widely from 30 to 250 mm. However, blasthole diameters ranging between 51 and 115 mm are commonly used for pre-split blasting in underground and surface excavations. ISEE (2011) recommended 51 to 89 mm whereas Olofsson (1998) recommended a hole diameter of 30 to 64 mm, Jimeno et al. (1995) recommended hole diameter of 35 to 75 mm and Hagan and Mercer (1983) recommended a hole diameter of 75 to 250 mm.

Spacing between pre-split holes is generally determined based on the dynamic tensile strength of the rock and the borehole pressure of the explosive generated by the decoupled charge. The spacing between the pre-split holes as given by Calder and Jackson (1981) is:

$$S \leq \frac{D_h \times (\sigma_t + P_{de})}{\sigma_t} \tag{4.20}$$

where,
$S =$ Hole spacing
$D_h =$ Hole diameter
$\sigma_t =$ Tensile strength of the rock
$P_{de} =$ Decoupled borehole pressure of the explosive charge

Borehole pressures generated by fully-coupled and decoupled explosive charges as given by Jimeno et al. (1995) are:

$$P_{bh} = 228 \times 10^{-6} \times \rho_e \times \frac{VOD^2}{1 + 0.8\rho_e} \qquad \text{(Fully-coupled charge)} \tag{4.21}$$

$$P_{de} = P_{bh} \times \left[\frac{V_e}{V_b}\right]^{1.2} = P_{bh} \times \left[\frac{D_e}{D_h}\sqrt{C_l}\right]^{2.4} \qquad \text{(Decoupled charge)} \tag{4.22}$$

where,
$P_{bh} =$ Borehole pressure of fully-coupled charge (MPa)
$P_{de} =$ Borehole pressure of decoupled charge (MPa)
V_e & $V_b =$ Volumes of explosive and blasthole respectively
$\rho_e =$ Density of explosives (g/cc)
$VOD =$ Detonation velocity of explosive (m/s)
D_e & $D_h =$ Diameters of explosive and blasthole respectively
$C_l =$ Percentage of explosive column that is loaded

Table 4.54 Blast design parameters for pre-split given by different researchers.

	Blasthole diameter	Spacing	Linear charge concentration
Gustafsson	64 mm	0.6–0.8 m	0.46 kg/m
Persson	80 mm	0.6–0.8 m	0.57 kg/m
Sandvik Co.	102 mm	0.8–1.1 m	0.90 kg/m
Atlas Powder Co.	102 mm	0.9–1.2 m	0.89 kg/m
Blaster's Handbook	102 mm	0.6–1.2 m	0.38–1.12 kg/m
Hagan	115 mm	1.2 m	1.10 kg/m

Chiappetta (2001) used the following equation for the determination of borehole pressure generated by a decoupled explosive charge:

$$P_b = 1.25 \times 10^{-4} \times \rho \times (VOD)^2 \left[\frac{r_e}{r_b}\right]^{2.6} \tag{4.23}$$

where,
P_b = Borehole pressure of decoupled charge (MPa)
ρ_e = Density of explosives (g/cc)
VOD = Detonation velocity of explosive (m/s)
r_e & r_b = Radius of explosive and blasthole respectively

The explosive charge concentration required for pre-split blasting is generally determined based on the blasthole diameter. The empirical equations developed by Persson et al. (1994) and Jimeno et al. (1995) for the linear charge concentrations are:

$$Q_l = 90 \times d^2 \qquad \text{(Persson et al., 1994)} \tag{4.24}$$

$$Q_l = 8.5 \times 10^{-5} \times D^2 \quad \text{(Jimeno et al., 1995)} \tag{4.25}$$

where,
Q_l = linear charge concentration (kg/m)
d = blasthole diameter (m)
D = blasthole diameter (mm)

The blast design parameters recommended by different researchers for pre-split blasting are given in Table 4.54.

4.5.4.1   Blast design patterns of pre-split blasting

The proposed blast design patterns for pre-split blasting at the T-3 area of Sharda highwall mining project were made based on the geo-mechanical properties of rocks and the nature of deposits. The blasthole diameter used for production blasts at the project site is 160 mm. However, smaller blasthole diameters would be preferred for

Table 4.55 Borehole pressure produced by different explosives in a 115 mm blasthole diameter.

Explosive type	Diameter	Density	VOD	Borehole Pressure
Non-Permitted	25 mm	1.1 g/cc	3800–4000 m/s	42.92 MPa
P-5 (Emulsion)	32 mm	1.1 g/cc	3500–4000 m/s	69.61 MPa
P-5 (Slurry)	32 mm	1.1 g/cc	3400–3800 m/s	60.64 MPa

pre-split blasting. Hence blast design patterns for both 115 mm and 160 mm diameter were proposed for the experimental blasts. The drill machine should be able to drill with great accuracy with minimal hole deviation up to the height of at least 14 to 15 m. For better positioning of the drilling rig, a crawler mounted drill machine was recommended.

4.5.4.2   Pre-split blasting patterns using 115 mm blasthole diameter

The type of explosive proposed for pre-split blasting to obtain a decoupled charge was of 25 mm diameter, 125 gm weight per cartridge having detonation velocity less than 4000 m/s. However, the availability of specially designed explosives for pre-split blasting in the market might also be explored. Similarly, the possibility of using permitted explosives with 32 mm diameter could also be assessed as permitted explosives of lower strength could be easily available in the surrounding underground coal mines of SECL.

Considering the equation developed by Chiappetta (2001), the borehole pressures produced by 25 mm diameter of non-permitted explosives, 32 mm permitted (P-5 type) emulsion and slurry explosives in 115 mm blasthole diameter are given in Table 4.55.

The study carried out at the site indicated that coarse to very coarse-grained sandstone of white colour constituted the majority of the rock mass. The uniaxial compressive strength of coarse-grained sandstone varied from 4.0 to 29.16 MPa and very coarse-grained sandstone varied from 2.02 to 18.59 MPa. The mean values of the compressive strength for coarse-grained and very coarse-grained sandstones were 9.02 MPa and 12.92 MPa respectively. Considering the dynamic *in situ* compressive strength of rock as two times its static strength, the dynamic compressive strength of coarse-grained sandstone would vary from 8 to 58 MPa with a mean value of 26 MPa. Similarly, the dynamic compressive strength of very coarse-grained sandstone varied from 4 to 37 MPa with a mean value of 18 MPa.

The borehole pressures produced by emulsion- and slurry-based permitted explosives (P-5 type) of 32 mm diameter in a 115 mm blasthole diameter were 69.61 MPa and 60.64 MPa respectively. However, these values are much higher than the dynamic compressive strength of rocks. The borehole pressure generated in a 115 mm blasthole diameter by a 25 mm diameter of non-permitted explosive was 42.92 MPa. This value was less than the borehole pressure produced by 32 mm permitted explosives. Therefore, a 25 mm diameter was preferred for pre-split blasting. However, the mean dynamic compressive strengths of coarse-grained sandstone and very coarse-grained sandstone were still less than the borehole pressure. The linear charge concentration would therefore be reduced in order to prevent excessive crushing.

For the determination of the spacing between pre-split holes, the mean dynamic tensile strength of coarse-grained sandstone was used. The mean value of the tensile strength is 1.52 MPa. The dynamic tensile strength can be taken as 4.56 MPa. The spacing of holes for a pre-split line can be calculated as:

$$S \leq \frac{D_h \times (\sigma_t + P_{de})}{\sigma_t} = \frac{0.115 \times (4.56 + 42.92)}{4.56} = 1.2 \, \text{m} \tag{4.26}$$

Therefore, the spacing between pre-split holes became 1.2 m. The linear charge concentration based on the empirical equations given by Persson et al. (1994) and Jimeno et al. (1995) for a 115 mm blasthole diameter were:

$$Q_l = 90 \times 0.115^2 = 1.2 \, \text{kg/m} \qquad \text{(Persson et al., 1994)} \tag{4.27}$$

$$Q_l = 8.5 \times 10^{-5} \times 115^2 = 1.12 \, \text{kg/m} \qquad \text{(Jimeno et al., 1995)} \tag{4.28}$$

Based on the above calculations, the linear charge concentration for a 115 mm blasthole diameter became 1.12 to 1.2 kg. However, the borehole pressure generated by a continuous explosive charge of a 25 mm diameter was higher than the mean dynamic compressive strength of rocks. Therefore, a linear charge concentration of 0.85 to 1.0 kg/m had been proposed for the trial blasts. For the charging of explosives in pre-split holes, detonating fuse having a core charge of 10 g of PETN/m could be used. The proposed blast design pattern for 115 mm blasthole diameter is given in Figure 4.89.

4.5.4.3 Pre-split blasting patterns using 160 mm blasthole diameter

The borehole pressure generated by different types of explosives in a 160 mm blasthole diameter is given in Table 4.56. Due to higher decoupling ratio, the borehole pressure produced by different explosives was lower in the case of a 160 mm blasthole diameter when compared to a 115 mm blasthole diameter. Therefore, either a 25 mm diameter of non-permitted explosive or a 32 mm diameter of permitted explosive (P-5) could be used for a 160 mm blasthole diameter (Figure 4.89).

Considering a dynamic tensile strength of rock as 4.56 MPa, the spacing values for the pre-split holes for a 160 mm blasthole diameter came to:

$$S \leq \frac{D_h \times (\sigma_t + P_{de})}{\sigma_t} = \frac{0.160 \times (4.56 + 17.63)}{4.56} = 0.78 \, \text{m} \approx 0.8 \, \text{m} \tag{4.29}$$

(for 25 mm diameter non-permitted explosive)

$$S \leq \frac{D_h \times (\sigma_t + P_{de})}{\sigma_t} = \frac{0.160 \times (4.56 + 25.65)}{4.56} = 1.06 \, \text{m} \approx 1.1 \, \text{m} \tag{4.30}$$

(for 32 mm diameter slurry P-5 type explosive)

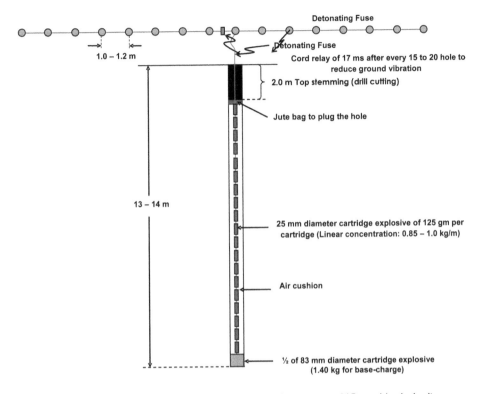

Figure 4.89 Proposed blast design pattern of pre-split blasting using 115 mm blasthole diameter.

Table 4.56 Borehole pressure produced by different explosive types in a 160 mm blasthole diameter.

Explosive type	Diameter	Density	VOD	Borehole Pressure
Non-Permitted	25 mm	1.1 g/cc	3800–4000 m/s	17.63 MPa
P-5 (Emulsion)	32 mm	1.1 g/cc	3500–4000 m/s	29.44 MPa
P-5 (Slurry)	32 mm	1.1 g/cc	3400–3800 m/s	25.65 MPa

The required linear charge concentration based on the empirical equations prescribed by Persson et al. (1994) and Jimeno et al. (1995) for a 160 mm blasthole diameter were:

$$Q_l = 90 \times 0.160^2 = 2.3 \, \text{kg/m} \qquad \text{(Persson et al., 1994)} \qquad (4.31)$$

$$Q_l = 8.5 \times 10^{-5} \times 160^2 = 2.18 \, \text{kg/m} \qquad \text{(Jimeno et al., 1995)} \qquad (4.32)$$

The required linear charge concentration for a 160 mm blasthole diameter was higher in comparison to a 115 mm diameter. Therefore, a 32 mm diameter cartridged

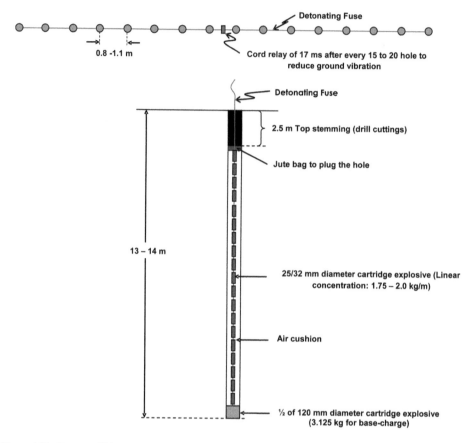

Figure 4.90 Proposed blast design pattern of pre-split blasting for experimental blasts using a 160 mm blasthole diameter.

explosive would be more preferable for charging of the hole in case of pre-split blasting with a 160 mm blasthole diameter. The proposed blast design patterns for pre-split blasting using a 160 mm blasthole diameter are given in Figure 4.90.

4.5.4.4 Important notes

The blast design patterns for pre-split blasting were made for 115 mm and 160 mm blasthole diameters. The following parameters were optimised during day-to-day operations:

a) spacing of pre-split holes
b) linear charge concentration of explosives
c) quantity of base charge
d) top stemming length
e) distance between buffer holes and pre-split holes
f) ground vibration and noise/generated from pre-split blasts

4.6 PERFORMANCE ANALYSIS OF HIGHWALL MINING AT OCP-II AND MOCP

ADDCAR-make continuous highwall miners were employed in the OCP-II and Medapalli OCP of SCCL for highwall mining operations. During the assessment period, in the OCP-II, only two seams (seam I and seam II) were mined out. The length of the highwall available for mining was divided in two panels namely panel A and panel B. In seam I both the panels were extracted by highwall mining operations while in seam II only panel B was extracted. Panel A extractions in OCP-II were aligned along an apparent dip direction, whereas in panel B they were along the full dip. The mining operations at OCP-II were completed successfully in November 2011 where the maximum monthly production achieved was in the range of 40,000 to 45,000 tonnes.

In MOCP, a total of five seams were exposed at the highwall. The length of the highwall available for mining was divided in two panels namely panel A and panel B. At MOCP, panel A extractions were almost along the strike direction with some up-dip gradient, whereas panel B was aligned along a down-gradient apparent dip. During the investigation time, mining operations were started in panel A and by the end of September 2012 the extraction of coal from seam IV and V was completed. Extraction from panel A seam III was in progress. An average monthly production of about 25,000 tonnes was achieved at MOCP.

The occurrence of roof falls, minor faults, exposure of stone band in the seam, steepness of the seam, working in cross gradient, and poor power supply etc., were some of the issues that had jeopardised the full production capacity of the CHM. The experiences at these two sites are described in the subsequent sections.

4.6.1 Highwall mining at OCP-II

Panel A

In panel A of seam I, a total of 18 drivages/web cuts were made (Figure 4.91). A few of them could not be driven to the full depth either because of seepage of water or roof fall. Apart from these, few web cuts were not extracted because of an unfavourable gradient for machine working. Overall 76% hole completions were achieved. An ADDCAR sequence problem, damaged inclinometer cable due to rock fall, sudden reduction in seam thickness, damage to cutter motor clutch, belt jamming, conveyor thermal tripping, roof fall and overloading of belt were the issues that stopped machine operations.

Panel B

In seam I, 37 web cuts were driven successfully and the highest hole completion ratio of 90% was achieved as shown in Figure 4.92. Panel B of seam I had the most favourable geological conditions, though the seam gradient was slightly steeper; a monthly production of about 40,000–45,000 tonnes was achieved during this extraction.

In seam II, except for 2–3 web cuts out of 39, none could achieve the designed depth of penetration, mainly on account of the frequent fall of the clay roof and the encounter of hard stone bands during coal cutting. Only 40–45% hole completion was

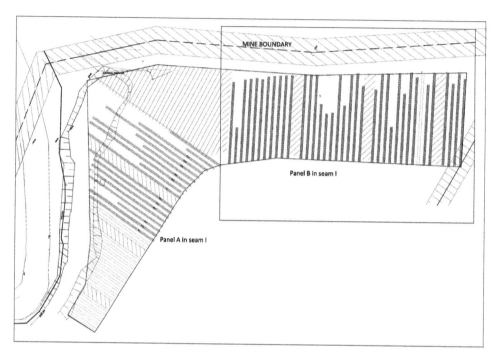

Figure 4.91 Plan view of penetration achieved in panel A and B of seam I of OCP-II.

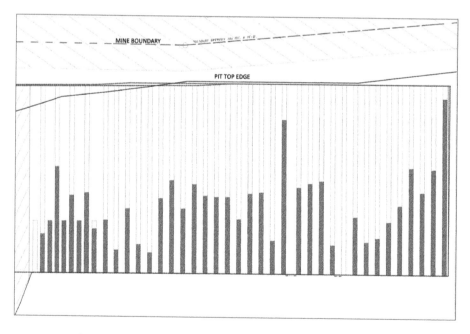

Figure 4.92 Plan view of penetration achieved in seam II of OCP-II.

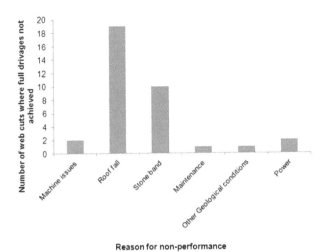

Figure 4.93 Factors affecting achievement of designed drivage depths in seam II of OCP-II.

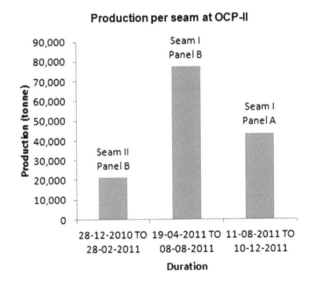

Figure 4.94 Production per seam at OCP-II during execution.

achieved (Figure 4.92). As this was the first panel to be extracted by highwall mining in India, the crew also needed a few months to get trained.

Based on available records, a chart was prepared showing causes of machine stoppage against the number of web cuts that could not achieve full penetration, see Figure 4.93. It was found that roof fall and the occurrence of a stone band restricted drivages to the full depth in almost 29 web cuts. Production per seam at OCP-II is given in Figure 4.94.

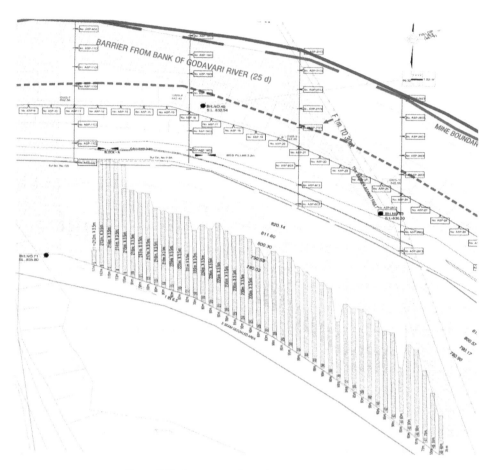

Figure 4.95 Recovery achieved in seam V at MOCP.

4.6.2 Highwall mining at MOCP

Panel A

A total of 45 web cuts were made in seam V and none of them was driven to the full designed depth as the coal turned to hard clay in many places. The seam consisted mostly of clay after a distance of 100 m from the highwall. Apart from this, due to working in a cross gradient, the extraction thickness of the comparatively thinner seam got reduced by about 0.5 m. Thus, the average hole completion percentage for seam V was as low as 39% (Figure 4.95).

Similar to seam V, extraction of seam IV too faced interruptions many a time, mainly due to the existence of hard stone bands within the seam. In almost every drivage, hard stone bands were encountered at different depths of penetration, making part of the seam irrecoverable. Such stone band occurrences are shown in Figures 4.96 and 4.97. In total 45 drivages were made and only a 65% hole completion ratio was achieved (Figure 4.98).

Figure 4.96 Stone band intrusion in coal seam at MOCP.

Figure 4.97 Occurrences of stone band at MOCP.

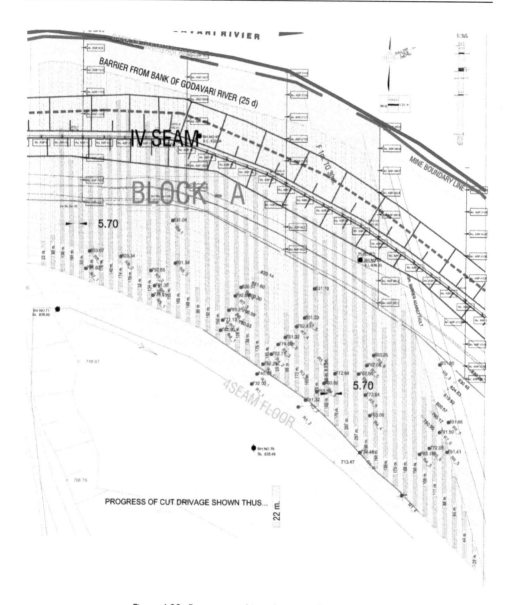

Figure 4.98 Recovery achieved in seam IV at MOCP.

During the period of study, extraction was in progress in seam III. The 6.0 m thick seam was being extracted using a two-pass extraction. In this seam too, extraction work was hampered many a time by adverse geological conditions. Coal in this seam was very hard with shale and stone bands. The composition of the immediate roof was of shale, about 1 m thick. The cutting of hard coal and often the shale roof led to the early consumption of picks and thereby loss of working time due to the need for frequent replacements. Similarly, the shale roof was prone to roof falls in large slabs.

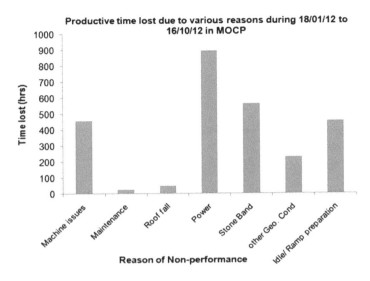

Figure 4.99 Productive time lost due to various issues at MOCP.

As the highwall miner cut the roof, this led to the frequent withdrawal of the machine for the removal of the shale slabs.

To show the varying recovery percentage in different seams and panels, an assessment of breakdown time was made for the MOCP. A chart showing the reason of breakdown and the time lost in hours was prepared for the period from 18-01-2012 to 16-10-2012 and is shown in Figure 4.99. It was found that power issues alone contributed to major production loss and amounted to a loss of approximately 896 hrs. Stone band occurrences, other geological conditions and roof fall together amount to a loss of 848 hrs during a time period of 10.5 months. Apart from that, other machine issues and ramp preparation had also consumed much of the productive time.

Monthly production of the MOCP highwall mining operations for the period 18 January 2012 to 16 October 2012 is shown in Figure 4.100. From 16 April 2012 to 20 May 2012 production was stopped either due to ramp preparation, repair of transformer or due to machine maintenance work. Similarly, the cutter machine motor got overloaded due to the cutting of the hard stone band and was broken down during 28 August 2012 to 12 September 2012. Similarly, there was an electrical breakdown of the machine during the first week of October 2012 due to voltage imbalance.

4.6.3 Amenability to highwall mining

Highwall mining technology had been newly introduced in India, the two sites mentioned being the initial experiences. The true amenability of highwall mining operations in Indian coalfields are yet to be assessed. There are many factors which may influence a particular site or a seam for its prospects for highwall mining. The productivity of the highwall miner also depends on the amenability of the coal seams. The important factors contributing to the amenability as well as productivity of a highwall miner

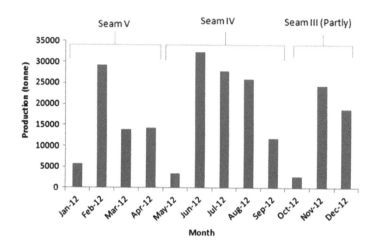

Figure 4.100 Monthly production at MOCP.

include the geological conditions, availability of spares and repairing facilities, supply of power, work culture and human resources.

A pre-assessment of the degree of amenability for highwall mining can give an idea of the expected production rate. As the cost of mining per ton of coal in highwall mining is linked to the rate of production, it will help the mining companies to estimate the production cost prior to the commencement of mining. It will also help in making improvements and maximise the rate of production at a given site. These initial case studies of highwall mining experiences at SCCL mines in India would help in assessing individual seams and the chosen sites for their highwall mining amenability and to come out with anticipated production rate targets. It should also be noted that in Indian conditions, other mechanised mining systems such as road headers, longwall mining, continuous miners, etc. have also recorded below par performances compared to those operating in other parts of the world. Considering these figures, the highwall miner operating at SCCL had been delivering a reasonable performance even in such difficult geo-mining conditions.

4.6.4   Summary of findings

From the analysis of the highwall mining performance and productivity so far recorded at OCP-II and MOCP of SCCL, the following conclusions were drawn:

- The geo-technical conditions were more adverse at these two sites in comparison to those seen in the USA and Australia so far.
- The adverse geology at the SCCL sites included a steeper seam gradient, poor roof conditions (wherever clay was encountered in the roof), harder coal seams, stone bands within the seams, minor faults, etc.
- Interruptions in machine operations were found to occur, directly or indirectly, mostly due to these adverse geological conditions.

- The other major factor found which delayed the machine performance was the poor quality of power supply, which has now been rectified to some extent. Other machine issues such as availability of faster repair and replacement facilities were also found to affect the overall productivity of the machine. However, these issues could be improved with experience over a period of time.
- Coal seams with shale/stone bands within the planned extraction section were very hard for the cutters, causing faster pick wear and overloading of the cutter motors.
- However, it was encouraging that even in such adverse geological conditions the machine had done a decent job of extracting coal at a reasonable rate, with the cost of production well within the average market price of Indian coals. It should be noted that this coal was otherwise un-mineable by any other means, and was considered lost under the highwall slope.

4.7 CONCLUSIONS

Although the key to successful operations lay in the scientific design of web pillars, increased recovery of locked-up coal had been the main requirement at all the highwall mining sites. Design methodologies carried out by CSIR-CIMFR included estimation of long and narrow pillar strength through an empirical equation developed by the institute (Sheorey, 1997) with modifications and application of numerical simulation to consider multi-seam interactions, pillar stress, etc. The highwall mining design methodology and performance analyses at all the sites have been discussed in detail in this chapter. Mining operations at all the sites were reasonably successful, performance analysis revealed that the occurrence of roof fall, breakdown on account of geological disturbances and poor power supply, exposure of stone bands in the seam, working in cross gradient, etc., were some of the issues that had jeopardised the full production capacity of the CHM.

It is already established that highwall mining technology can be a proven way to enhance the energy need of India and reduce the demand–supply gap of the country to a considerable extent. It promises greater safety to miners as compared to underground mines. Through this technology, the life of open-pit mines can also be extended after reaching lease boundaries without disturbing surface dwellings, while maintaining a good economy and productivity. Highwall mining technology application in India is therefore seen as a boon to recover coal from Indian opencast mines that are reaching their pit limits and also to extract coal from thin seams which are otherwise uneconomical. In many cases, the existence of surface dwellings which prohibited the expansion of currently running open-pit mines can be further continued through the implementation of highwall mining.

It is necessary to pre-assess the machine performance and anticipate the production rate on a site-to-site basis prior to planning the highwall mining operations, as the production rate and amenability to highwall mining is directly linked to the cost of production. Efforts are underway with foreign collaboration to frame guidelines and norms as well as assessment of highwall mining amenability for Indian conditions.

Chapter 5

Structural mapping

5.1 INTRODUCTION

Highwall mining could be a high risk activity in cases where the highwall slope fails, burying mine workers and machinery. In some cases, the highwall slopes accommodating the highwall excavations might not fail, but if the roof collapsed within the highwall excavation, induced by local failure due to faults, joints, fractures and web pillar failure, it could bury the continuous miner and conveying system in the excavations. In that case, the mine would virtually lose its revenue by losing expensive machinery. However, as highwall mining is operated remotely and no mine worker is allowed inside the excavation entries, the technology is considered to be a safe technology with respect to the safety of men. The highwall mining technology could be dangerous if the highwall slopes fail during its operations, as this has direct consequences for the safety of mine workers and also economic implications. Further, detachment or instability of rock blocks due to stress relief at the face, blasting or the formation of local wedges (i.e. local failure, partial face failure, full face failure, etc.) could affect the safety of the workers. So, this chapter is aimed at developing a framework for structural mapping, modelling and slope failure analysis with a view to assessing highwall slope hazards and their remedial measures.

The recently published guidelines for open-pit design present a detailed framework for the management of uncertainty and risk in open-pit slope design throughout the conceptual, pre-feasibility, design, and operation phases (Read and Stacey, 2009). It outlines the levels of geotechnical effort and the target of data confidence for each project stage. ACARP project C19026 of CSIRO provides improved hazard management strategies for mitigating rock-fall hazards (Giacomini et al., 2012) in coal mines.

The work described in this chapter builds upon past research and integrates with present work into structurally controlled hazards within highwall environments. Previous ACARP research undertaken by the then CSIRO Division of Exploration and Mining (Poropat et al., 2004) established the basic computational geometry algorithms necessary for the polyhedral modelling used for delineating exposed and internal rock blocks. Work supported by the Large Open Pit Mine Slope Stability Project (http://www.lop.csiro.au/) developed an application for the use of statistically sound fracture network generation coupled with polyhedral modelling for hazard detection in open cut mining environments. The general procedure developed by CSIRO using Sirovision™ software for structural mapping and modelling is depicted for a case example of the highwall at RamagundamOpencast Project-II (OCP-II) of M/s Singareni Collieries Company Limited (SCCL) (AISRF Final Report: Agreement ST050173, 2015).

5.2 FRAMEWORK FOR STRUCTURAL MAPPING AND MODELLING

This section provides a framework for structural mapping and the first pass stability analysis of slopes using Sirovision^{TM} software. The general framework for structural mapping and modelling would generally incorporate the following steps:

1   Define the objectives of the stability analysis;
2   Gathering data, including images acquired in the field;
3   Use of software such as Sirovision^{TM} to generate 3D images from the images collected in the field;
4   Digital mapping of joints using Sirovision^{TM}, distribution analysis and statistics;
5   Performing a first pass stability analysis using discontinuity data with a stability modelling software such as Siromodel.

The software packages Sirovision^{TM} and Siromodel will be referred to several times in this chapter. The Sirovision^{TM} system has been developed by CSIRO and is used to measure and analyse discontinuities in a rock mass. This system is based on a photogrammetric technique that allows 3D images to be constructed from 2D images captured with a digital camera. Structures can be mapped from these images quickly and safely, with no need to approach a potentially unstable highwall.

The Siromodel system has also been developed by CSIRO as one part of Stream 2 of the Large Open Pit Mine Slope Stability Project. It allows first pass stability analyses to be performed based on discrete fracture network representations of the discontinuities in a rock mass.

5.3 DIGITAL MAPPING AND ANALYSIS

There are several systems available on the market that allow digital mapping of a rock mass structure. Sirovision^{TM} is a structural mapping and analysis system that takes images captured using digital cameras and turns them into high-precision 3D images of the rock mass surface. These 3D images can then be used to map and analyse the distribution of joints daylighting on the surface of the rock mass. This allows large areas to be mapped quickly and safely as personnel are not required to approach hazardous slopes. The system has both open cut and underground implementations.

5.3.1 Digital photogrammetry

The large, quasi-planar exposures associated with highwall mining provide an ideal environment for structural mapping, particularly when performed via remote, digital means. Digital photogrammetry allows the rapid acquisition of georeferenced data with high spatial resolution and colour representation to be acquired from a safe working distance. Digital mapping allows the practitioner to map much larger areas of exposure than were typically possible via manual methods. Combined, the two technologies allow for the creation and collation of voluminous data sets of rock mass structures within a practical time. This in turn provides a way of quantifying uncertainties associated with structural properties. The availability of data from larger sampling areas than those traditionally used can reveal larger uncertainties than are sometimes assumed if only small sample sizes are used.

A general rule of thumb regarding exposure mapping is to include approximately one hundred to two hundred structures per joint set (ISRM, 1978). There is a significant amount of work regarding the need to perform bias correction on the mapping of trace data, which is especially significant for trace lengths (Mauldon, 1998). Borehole data suffers from similar issues as data acquired using scan lines. In general, failure to bias-correct data, particularly when a small number of structures are available, can introduce considerable errors to the estimation of the fracture frequency and trace length. The latter is directly related to the size of the fractures and therefore greatly influences predictions of fracture connectivity, rock mass fragmentation, block size distribution and slope stability analyses. For severely censored data, under-estimates of the mean trace length will have obvious consequences regarding the propagation of uncertainties into the resulting geotechnical analyses.

Sirovision™ includes leading edge photogrammetric software to allow the creation of 3D images from pairs of 2D images. The software can produce spatial data with accuracies of the order of better than 1 part in 10,000 when used under controlled conditions, but more typically 1 part in several thousand when used with typical field procedures.

All that is required to acquire good quality photographs in an open-cut environment is the use of a good quality digital SLR camera with at least a 6 megapixel resolution with an appropriate lens. The lens focal length chosen will depend upon the distance to the rock face and the accuracy required by the user. Sirovision™ provides planning tools to define field procedures and choose appropriate lenses. There is always a trade-off between resolution and area covered, with greater coverage giving a lower resolution for a given camera resolution. Additional accessories such as a tripod, levelling plate, compass and laser range finder can aid in the acquisition and provide extra precision if required.

The first step is the acquisition of the images. The field procedure used will depend on the preferred method of georeferencing to be adopted. The Sirovision™ manual describes the following in summarised forms:

- The camera location and shooting direction (azimuth, dip and tilt) with a single control point on the face can be used.
- Alternatively, a minimum of three control points in the field of view, positioned so as to form an equilateral triangle, can be used.
- Alternatively, a minimum of three camera positions, positioned so as to form a non-linear alignment, can be used.

In summary, field procedures comprise of the following steps:

1   Select the area of wall to photograph;
2   Select the camera location – the distance to the wall will be determined by the focal length of the lens being used and the resolution required. The distance between

camera locations for left and right-hand images should typically be 1/5th to 1/7th of the distance to the face. Another important consideration is the angle to the face in the horizontal and vertical planes. Ideally, the camera should be perpendicular to the rock face, but this is rarely possible in practice;

3 Depending on the chosen field procedure, the placement of control points in the field of view may be required;

4 Take left-hand and right-hand photographs of the same area of the rock face to generate the images needed to create a 3D image;

5 Depending on the chosen field procedure, the first camera position may require marking for survey purposes (e.g. using a traffic cone or similar marker);

6 Repeat for adjacent sections of the highwall. It is recommended that one should use an overlap of around 25 to 30% of adjacent sections to ensure accurate 3D image alignment for mosaics.

When image acquisition is complete, surveying of the marked camera positions and/or control points can be conducted. The required accuracy of the surveying will depend on the type of analysis being performed. The Sirovision™ documentation provides guidance on this point. Once the images are collected in the field and the survey data obtained, the 3D images can be created. Once images are acquired, the main stages of the processing for creation of georeferenced mosaics are:

3D image creation;
3D image alignment for mosaic creation; and
georeferencing.

5.3.2 Georeferencing

For kinematic and stability analyses, accuracy in estimating relative orientations of structure sets and highwall geometry are more important than accuracy in georeferencing. Hence the field procedures typically adopted have a good trade-off between time required and georeferencing error. Alternate procedures could be used if required, which provide extremely high levels of georeferencing precision. In general, proper planning of field work is required to ensure the georeferencing accuracy is consistent with the types of analysis being conducted.

At each mine site, local surveyor support is usually provided for surveying ground control points and camera locations. Typical survey point adjustment errors, as reported during the georeferencing process, were around 200–700 mm, sufficient for most purposes.

5.3.3 Structure mapping

Once the 3D mosaic is georeferenced, digital mapping of structures such as joints, faults, bedding, etc. can be undertaken. As with traditional field-based mapping, the practitioner should be aware of certain issues and biases, which are discussed below.

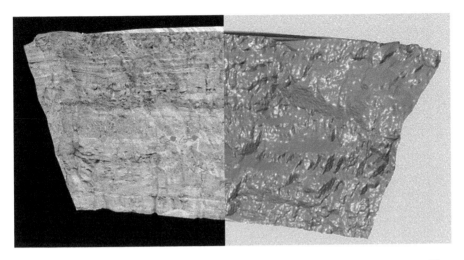

Figure 5.1 Texture mapped (left) and false colour (right) of a 3D model using Sirovision™.

5.3.3.1 *Curvature*

All rock mass defects have curvature or waviness of varying degrees. The significance of the curvature is dependent upon the structure size and is clearly important for large defects such as faults, but may also be significant for geometrical analyses of minor structures. It is presumed that a component of the variation in orientation for a joint set is attributable to the simplifying assumption that planes can be fitted to these defects. It is also to be noted that if exaggerated curvature is evident, the defect should be mapped at several elevations, meaning the structure count will be artificially increased. There is as yet no established method to digitally map and record curvature on an individual structure.

5.3.4 Lighting and contrast

One of the benefits of 3D photogrammetry is the automatic registration between the digital image and the underlying point cloud. However, this use of 'texture-mapping' for 3D visualisation can mask the true quality of the 3D image data. Further, for sedimentary geology, the large number of bedding planes can make delineation of smaller joint structures difficult in certain lighting conditions.

The use of artificial lighting and surface colouring can alleviate this. This involves using a computer generated light source and an arbitrarily coloured surface mesh employing specular reflection. Figure 5.1 shows an example of a traditional texture map and artificial lighting visualisations of a 3D model using Sirovision™. By removing the colour information pertaining to the bedding layers and applying an artificial colour to the surface, the artificial lighting highlights the underlying structure more clearly in this example.

To gain all the effectiveness of the artificial lighting effect, one should use the digital image regularly to ensure that the structures being mapped correspond to those

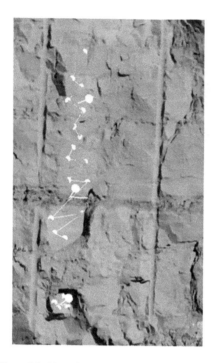

Figure 5.2 Use of bridging tool to join two separately mapped defects (left) into one (right).

of interest (e.g. joints) rather than others (e.g. pre-split barrels, bucket marks, post scaling, etc.).

5.3.5 Estimation of structure extent

The mapping process tends to focus the eye on structures with similar characteristics (e.g. orientation and size). Upon reviewing the results, one may realise that multiple structures actually form part of a single defect (see Figure 5.2 for an example).

The bridging tool available in Sirovision^{TM} is valuable in this regard. It allows the user to select multiple traces for joining into a single trace. Defects associated with aggressively curved or deformed structures, or structures daylighting on either side of intermediate bedding layers, should not be bridged (e.g. Figure 5.3) but rather retained in the analysis as separate structures. This has the undesirable effect of increasing the structure count and decreasing the average structure size, however, variability in structure orientation is more correctly reported.

5.3.6 Measurement bias

Mapping of larger, deterministic structures, such as bedding planes and faults, is relatively straightforward. For the mapping of joints and smaller structures to be modelled stochastically, the question of which size cut-off (i.e. truncation bias) to accept is ever

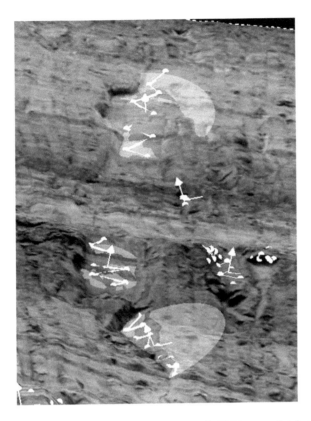

Figure 5.3 Example of multiple exposures of a highly curved defect.

present. This is partially controlled by the spatial resolution of the data, the quality of the images and the resources dedicated to the mapping process. One approach is to map structures down to a nominal size (e.g. 1 to 2 m) for a detailed analysis of a limited region and down to a few metres otherwise. This truncation may be tolerable if the focus is on stability analysis.

Another bias that is inherent to digital mapping, but less important in traditional methods, is resolution bias. Digital mapping potentially gives the practitioner access to the entire highwall. However, under normal field procedures, the imaging of the upper elevations will utilise the same camera positions as those used for the lower elevations, meaning the effective range can be much greater. If the resulting spatial resolution is not sufficient, accurate estimation of the trace orientation will be impossible, even for the cases where trace persistence can be measured (Figure 5.4). Resolution bias also affects the ability to map traces on quasi-planar surfaces (i.e. lacking topography) as opposed to exposed joint surfaces (i.e. planes). The practitioner must often rely on digital manipulation of the 3D image (zoom, pan, rotate) or the use of artificial lighting methods to provide sufficient contrast to map an exposed surface. However, if the defect daylights 'cleanly' on the surface of the wall with little or none of its plane being evident, then such methods will be ineffective.

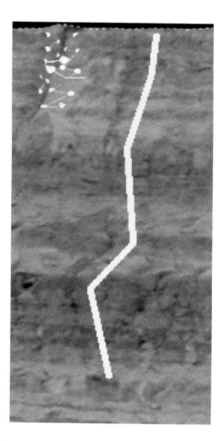

Figure 5.4 Example of a large defect at the top of a highwall (seen to the left of the yellow line).

Although the use of window mapping avoids the severe sampling bias associated with linear sampling techniques, the linear geometries of coal highwalls make orientation bias an issue (Park and West, 2002). Characterisation of the frequencies of structures sub-parallel to the wall is extremely difficult. Traditional methods of correcting for this bias are only useful if sufficient numbers of structures from the set have been observed to derive accurate set parameters (Terzaghi, 1965). Alternatively, if no quantification of frequency is possible, a conservative approach should be adopted where the structure should be incorporated in a scenario analysis. Orientation bias is a serious issue and provides a limit on the predictive capabilities of the methodology proposed in any research.

A similar bias also results from the imaging geometry used to image the highwall. If a structure set is unfavourably oriented, relative to the ambient lighting or camera position, it will be difficult to resolve and may even be absent from the spatial data if image matching has failed. No simple correction method is possible and, if evident, the wall should be reimaged with an alternate imaging geometry.

The use of digital mapping potentially provides much larger exposures for structural analysis than traditional means. This reduces the impact of censoring bias.

Another mitigating factor is the presence of severe structure termination against inter-mediate bedding layers. Fitting a distribution to trace lengths can yield fracture radius estimates. Sedimentary geology provides constraints that allow a more deterministic modelling of structure size. Using structural modelling software such as Siromodel provides a way to quantify this bias using instabilities (e.g. wedges) visible in the highwall.

Finally, the techniques adopted in this chapter attempt to mitigate the effect of the so-called size bias. The tendency for sampling methods to over represent larger structures can be compensated for in the fracture modelling process given certain assumptions (e.g. joints are circular discs). However, the validity of these assumptions should always be questioned.

5.4 BLAST DAMAGE

Depending on the requirements of the analysis, pre-split and blast-induced fractures may or may not be of interest. In any case, distinguishing between the blast damage and *in situ* joints is usually important and relies heavily on the ability of the practitioner to recognise the characteristics associated with blast damage, such as plumose structures, fracture orientations parallel to the highwall and proximity to blastholes. An example of obvious blast damage is shown in Figure 5.5. Cases where damage is not so obvious constitute another source of error, normally resulting in an over-estimation of joints parallel to the highwall. This problem may become more evident via examination of a stereonet after some structure mapping has taken place.

Methods to predict the fragmentation from blasting, given the *in situ* rock mass characteristics, blast geometry and explosive properties, have been studied (Hoek, 2012). The thickness of the blast-damaged zone is difficult to predict and will depend

Figure 5.5 An example of blast damage.

on a number of factors, including the design of the blast and particular geological conditions.

5.5 STRUCTURE ANALYSIS

The photogrammetry package may or may not incorporate a structure mapping and analysis functionality, in which case alternative software will be required. Sirovision™ incorporates a sophisticated structure mapping and analysis functionality, the details of which are discussed in this section.

The Sirovision™ interface allows the user to visualise 3D mosaics, stereographic plot and histogram of structure persistence (Figure 5.6). The analysis module has facilities to:

- Plot discontinuity distributions using equal angle and equal area projections of dips or normals;

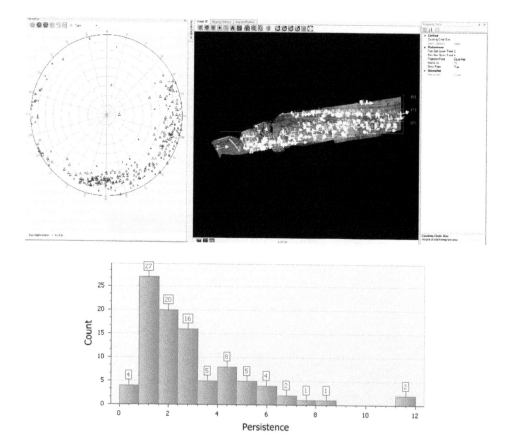

Figure 5.6 The Sirovision™ interface, showing structural mappings on a 3D mosaic of a coal highwall, stereographic plot and histogram of structure persistence.

- Plot discontinuity distributions using rose diagrams with the plot length or area proportional to the number of discontinuities;
- Analyse distributions using contouring or statistical clustering;
- Calculate normal set spacing for sets of discontinuities and visualise the discontinuities used for set spacing calculations;
- Visualise discontinuities and the 3D images or mine layout in which they exist; and
- Undertake a wedge based stability analysis using pairs of structures and user-defined intersection criteria.

Structures can be measured as a trace (e.g. the intersection of a joint with a surface) or as a plane (Figure 5.7). The plane can be mapped manually, by selecting points on the outline of a planar surface, or automatically. To automatically map a structure, the user can click on what they interpret to be a planar surface. Sirovision™ will then fit a plane (if possible) to that surface using parameters (minimum area and maximum variation in orientation) previously defined by the user.

Once the discontinuities in a 3D image have been mapped, they can be quickly analysed and grouped into sets based on orientation (automatically or manually) or domain (user-defined). In addition to the orientation of discontinuities, Sirovision™

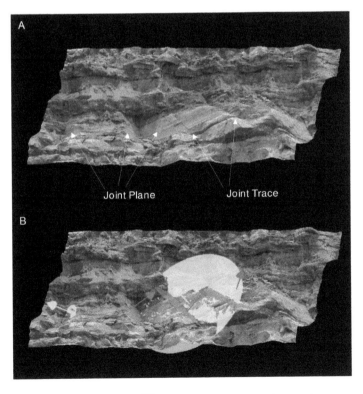

Figure 5.7 Structure mapping in Sirovision™. 'A' shows a 3D image with several structures highlighted. 'B' shows the same image with joints fitted.

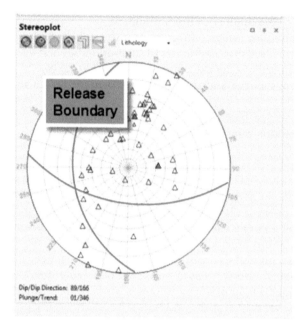

Figure 5.8 Visualisation of the results of a Sirovision™ stability analysis using either stereonet or wedge visualisation (Courtesy Sirovision™ user manual, CAE Mining, 2013).

calculates the discontinuity length (maximum chord length), area and spacing. An estimate of the persistence of a discontinuity is obtained using the maximum chord length of the plane.

The discontinuity area is calculated by drawing a disc where the persistence is the diameter. The area of that disc is then calculated to give the discontinuity area.

Sirovision™ calculates the normal set spacing for a joint set by determining the average normal for the set and estimating joint spacing along the normal.

The Stability Analysis tool allows one to select two analysis sets and analyse them with set geotechnical criteria to automatically detect potential unstable wedges. The analysis uses combinatorial analysis and user specified structure properties to identify pairs of planes that could potentially be associated with wedges in the highwall. The friction angle, cohesion, persistence and factor of safety threshold can be specified and the software will look for pairs of structures that could potentially form unstable wedges. The analysis results can be interpreted visually via stereonet or via the 3D visualisation (Figure 5.8).

5.5.1 Measurement accuracy

As previously stated, the dominant factor affecting the accuracy of measurements will be the quality of georeferencing applied to the 3D data. However, assuming perfect georeferencing, one can ask what limits the measurement accuracy?

Sirovision™ processes data from a range of sensors. The fundamental limit on the accuracy or precision of a measurement obtained using Sirovision™ is the accuracy, precision or resolution of the 3D images. There are limits on the performance of some algorithms, given the nature of real world data, however the dominant limits are from the images. In general terms, the orientation of planes can be measured in Sirovision™ with accuracies of +/−0.1 to 0.2 degrees at 100 metres for a 1 metre long feature. Sirovision™ provides three measures of quality for assessing the 'accuracy' of an estimate of the orientation of a plane:

- The Root Mean Square (RMS) deviation from the fitted plane in metres;
- The orientation standard deviation in degrees;
- The reliability of the orientation estimate.

It is important that these measures are understood, because the difference between what can be measured in the field and what can be measured from a 3D image may be significant. When the orientation of a surface is measured in the field, a spot measurement is taken. When a plane is selected in Sirovision™, a 'global' measurement is made across the whole area of that plane. If the plane is rough the difference between these may be significant.

5.6 STRUCTURAL MODELLING

This section outlines the methods for rapidly performing preliminary stability analysis of the highwalls in order to assist hazard identification for pit walls and highwall entries.

5.6.1 Discrete Fracture Network (DFN) modelling

The use of discrete geometric representations of fracture and joint networks to model the heterogeneity of a rock mass has greatly increased in recent years. The acceptance of this method has been a function of both the availability of software-based Discrete Fracture Network (DFN) generators, as well as the widespread availability of suitable computing resources. A DFN provides a more realistic simulation of rock mass heterogeneity than the more established Equivalent Continuum Medium (ECM) methods. ECM often relies on assumptions of parallel fractures within a given set, fracture ubiquity and assumed infinite persistence of fractures.

DFN techniques, therefore, better represent individual fracture properties, such as position, size and aperture. However, the potential of the DFN approach can only be realised if the geometry of the discontinuity network being generated is statistically similar to the *in situ* network of rock mass defects being simulated. The DFN should also honour the geological history of the structure formation by accurately representing the termination and truncation of the structures. If similitude is not achieved, the DFN approach can be considered, at best, to be an inefficient and overly complicated alternative to the ECM approach. At worst, the DFN can misrepresent the characteristics of the network, which has significant implications for the results from stability, fragmentation, hydrogeological and other analyses.

A DFN is developed using both the deterministic structures that have been explicitly measured (e.g. larger structures such as faults or large fractures that have been mapped on an exposure) and the stochastic or inferred structures. It is particularly important to include these large structures deterministically in a DFN because stochastic generation of large fractures can severely alter the geometries (and subsequent hydrological, stability and other analyses). The proposed methodology is to treat these larger structures separately to the stochastically generated ones and to resist the 'temptation' to bundle all structures into a single fracture 'family'. Uncertainties in these larger structures are typically different in nature (e.g. measurement uncertainty, model uncertainty). While stochastically generated structures have properties that are representative of the sample, deterministic structures retain large uncertainties, as each individual structure is modelled based on the measurements of a single structure. These uncertainties can be attached to the structures in what we have termed a quasi-stochastic analysis.

The DFN generation process for stochastic structures relies on the ability of the practitioner to create Probability Distribution Functions (PDF) representative of the fracture parameters, such as density, orientation and size. This can only be generated after the statistical parameters of the discontinuity matrix observed in the field have been quantified. Assumptions must also be made regarding the fracture shape (Dershowitz, 1984; Billaux et al., 1989).

A commonly adopted DFN model for the simulated rock mass is the Baecher model (Baecher and Lanney, 1978). This model assumes:

- Discontinuities are randomly distributed spatially (i.e. Poisson process);
- The sizes (persistence) of discontinuities can be described by log-normal or negative-exponential distributions; and
- The spacing between adjacent discontinuities from the same set (as measured by projecting discontinuity centroids along the mean set normal vector) can be described by negative-exponential distributions.

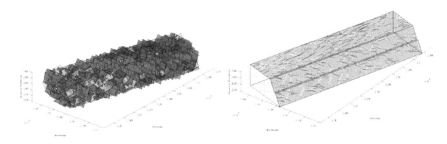

Figure 5.9 Example of a DFN realisation and resulting surface traces.

The assumption of a fracture model may only be adequate for generic studies. When the characteristics of a particular site are being investigated, the individual fracture properties must be estimated. These properties are represented as random variables and therefore statistical methods such as the Monte Carlo simulation must be used to generate simulated fracture networks (known as realisations), examples being shown in Figure 5.9.

5.6.2   Polyhedral modelling

Identification of the polyhedral or rock blocks present in a rock mass is an important tool in the application of block theory and numerical studies of rock mass stability. Polyhedral modelling does not provide the same level of physics sophistication as numerical stress-strain modelling, however, it is extremely useful for several reasons, such as the following:

- Compatible with fast, first pass analyses;
- More advanced than traditional analysis;
- Compatible with statistical sampling schemes such as Monte Carlo; and
- Able to provide estimates of *in situ* block size distribution, slope stability, rock mass fragmentation, etc., when coupled with accurate structural representations.

Used in isolation from numerical techniques, polyhedral modelling will provide an estimate of the 3-dimensional fragmentation of a rock mass but will not account for stress induced fracture propagation. Numerous polyhedral modellings have been described by various authors and used for the estimation of block size distribution (Wang et al., 2003; Rogers et al., 2007; Menéndez-Díaz et al., 2009). Traditionally, modellers required simplifications, such as cubic or hexahedral representations of the simulation volume; very small numbers of discontinuities; the assumption of planar topologies for the discontinuities; and the assumption of infinite persistence.

The other simplifications include a critical limit on the ability to represent a rock mass comprising of irregular shaped blocks or polyhedra. In particular, infinitely persistent discontinuities disallow the inclusion of concave extremities and can only yield convex polyhedral. These limitations are significant when assessing the stability of an excavation and *in situ* block size distribution (Elmouttie et al., 2010; Elmouttie and

Poropat, 2012). The polyhedral modeller utilised in this chapter represents an evolution of the modellers outlined in Lin et al. (1987); Jing and Stephansson (1994); Jing (2000) and Lu (2002) and is described in detail in Elmouttie et al. (2010). Examples of polyhedral models are shown in Figure 5.10.

5.6.3 Monte Carlo sampling

Any particular DFN realisation represents one of an infinite but equiprobable number that could have been generated, with large variances in properties being possible. Therefore, reliance on one or even a few realisations is statistically unsound, as analyses based on small numbers will be particularly sensitive to statistical outliers in the fracture geometry and properties. By repeatedly sampling from the probability distributions of the fracture parameters, generating a DFN realisation and finally performing a geotechnical analysis, one can capture the stochastic variability of the rock mass using multiple deterministic analyses. Figure 5.11 shows one hundred block size distribution curves based on one hundred DFN realisations. Raw curves are shown on the left, whereas mean and 95% confidence curves are shown on the right.

This data can be used to estimate mean parameter values and confidence intervals, which in turn can guide the practitioner in determining realistic outlier scenarios requiring more detailed, sophisticated and/or time-consuming analyses.

5.6.4 Analysis of coal highwalls using DFN approach

The use of DFN for the analysis of mining excavations has been largely limited to metalliferous and hard-rock operations. Some application to sedimentary and carbonate rock is evidenced in the literature (Dershowitz et al., 2007; Elmouttie et al., 2013). In bedding controlled geology, fracture sets have orientations that are sometimes controlled by the bedding orientation. Fracture intensity can also be related to bedding thickness. Wherever possible, critical bedding surfaces should be included as deterministic structures along with other major structures (such as faults). This is feasible with highwall mapping since the elevations, orientations and shapes of the bedding structures can be readily mapped, particularly using digital means. Fracture termination is also an important property to be captured in a DFN, particularly for sedimentary geology. Fracture size is greatly influenced by termination and this can greatly affect DFN properties such as connectivity, fragmentation and block stability (Elmouttie et al., 2009).

5.7 SIROMODEL

The Siromodel software application has been developed as one part of Stream 2 of the Large Open Pit Mine Slope Stability Project. The software allows the user to create a model of a section of an open-pit excavation for slope stability analyses. It also includes a multi-function Discrete Fracture Network (DFN) generator. This allows deterministic discontinuities, such as bedding planes and faults, and deterministic joints to be added to this model. Stochastically generated joints can also be produced upon each simulation by defining the statistical parameters for each set. Precedence settings can

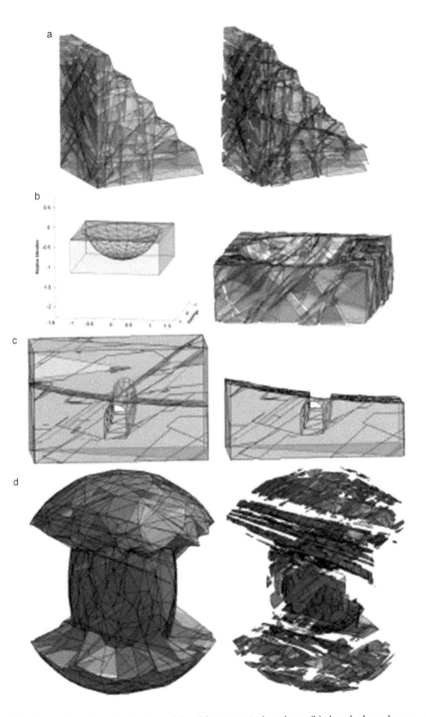

Figure 5.10 Examples of polyhedral models: (a) open-pit benches, (b) bowl shaped excavation, (c) underground excavation and (d) pillar (Elmouttie et al., 2010).

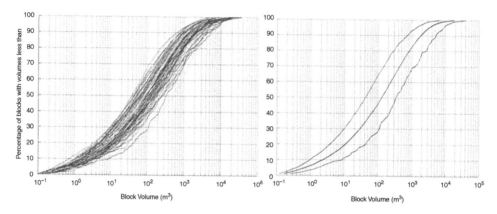

Figure 5.11 Example of predictions for block size distribution from 100 DFN realisations (Elmouttie et al., 2009).

be used to control the way various discontinuity sets terminate against each other. The software will determine the intersections of all these structures and the presence of all complete blocks within the model, known as a block model (also known as a poly-hedral model). Statistics on each block can be generated, including limit equilibrium stability assessment, and the blocks can be viewed individually or as an ensemble. A batch of simulations can be run constituting a Monte Carlo analysis, which can then be analysed.

The block finding algorithm used in Siromodel has been designed for handling finite persistence structures. It therefore does not utilise optimisations or efficiencies available to other algorithms that are limited to handling infinite persistent structures. If the structures being modelled consist almost entirely of persistent structures, then Siromodel may not be the optimal solution and one of the variety of spreadsheets or application based codes currently available may be more suitable.

5.8　CASE STUDY AT A HIGHWALL MINING SITE

5.8.1　Background

The case study uses previously acquired data at the end-user site of Ramagundam Opencast Project-II (OCP-II) of SCCL, but at a distance to the highwall mining blocks, where punch entries were driven (Guo et al., 2012; Huddlestone-Holmes et al., 2008; Karekal et al., 2008).

Figure 5.12 shows the general layout of mining operations. The red line shows the area that was imaged for the present study. The highwall mining blocks were up to the north. The highwall was around 125 m high in the vicinity of the punch entries (from the floor of I seam) and during the investigational period might have ended up significantly higher with OCP-II's continued expansion. It had a slope angle of approximately 35° to 40° and had been mined with a series of benches approximately

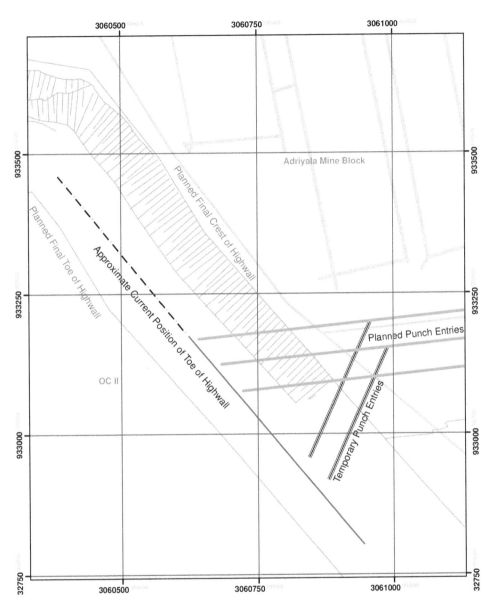

Figure 5.12 General layout of mining operations of OCP-II, SCCL *(CSIRO Report No-P2008/2412).*

10 m high. The benches had slope angles of up to 80° although they were typically much shallower. Benches at higher levels appeared to be eroding back to the overall slope angle so that the individual benches were no longer pronounced.

Table 5.1 details the joint and cleat patterns described in a study of slope stability for the OCP-II of SCCL (CSIR-CIMFR, 2005). This report looked at the overall stability of the highwall and spoil dumps. It recommended an ultimate pit slope angle of

Table 5.1 Joint and cleat sets in OCP-II, SCCL.

Joint/Cleat set	Dip direction	Dip	Spacing (m)	Persistence (m)
J1	130°	86°	1.5	2.5
J2	050°	14°	1.0	5.0
J3	232°	84°	1.5	1.0
CL1	050°	14°	0.2	0.6
CL2	155°	86°	0.1	1.0
CL3	260°	84°	0.4	0.6

47° with 10 m benches with slope angles of 80°. The report did not specifically look at the punch entries for the Adriyala project. The report recommended a well-developed drainage system and controlled blasting.

5.8.2   Image capture and processing

Mr. Phil Soole from CSIRO, during his visit to the Ramagundam area in July 2008, collected the images of the highwall. Thirty-four stereo pairs were collected over a distance of 500 m, from 3060960 mE, 9322805 mN in the southeast to 3060620 mE, 933170 mN in the northwest. Of these, 30 stereo pairs were processed into 3D images, comprising 12 "stacks"; the others were not required to provide full coverage.

Figure 5.13 shows the approximate camera locations for the 12 stacks. It is important to note that stacks 11 and 12 were more than double the distance from the highwall than the others. The images were collected using a Nikon D300 camera with a 10 Mega pixel resolution and a 60 mm lens. The distance to the toe of the highwall was approximately 120 m for images 1 to 10 and 300 metres for images 11 and 12 (Figure 5.13). The choice of camera locations was partly constrained by access to the site. All the images were corrected for lens distortion prior to processing. All camera locations were surveyed and control points were located along the toe and along the crest of the highwall. Survey data was provided by the SCCL management.

Three-dimensional images were located in real space using the two surveyed camera positions for each stereo pair and multiple control points. Some of the images did not contain control points and these were located relative to neighbouring images.

The 3D resolution of the images collected in Sirovision™, and the accuracy of the measurements made from them, was a function of the distance of the camera to the highwall, the resolution of the camera and the focal length of the lens used. Table 5.2 shows the accuracy and resolution of the images collected at OCP-II. This table shows the trade-off in accuracy for coverage for a given camera lens focal length. Images from stacks 11 and 12 covered significantly more area than the others. In image stacks 1 to 10, the accuracy of measured joint orientations was very good right down to 1 metre long joints. For images 11 and 12, this accuracy was diminished for 1 metre joints, particularly towards the top of the highwall, which was further from the camera due to the slope. However, this error was still acceptable.

The highwall at OCP-II had a moderate slope of 35° to 40° with numerous benches. This moderate slope combined with camera locations at the same elevation as the toe

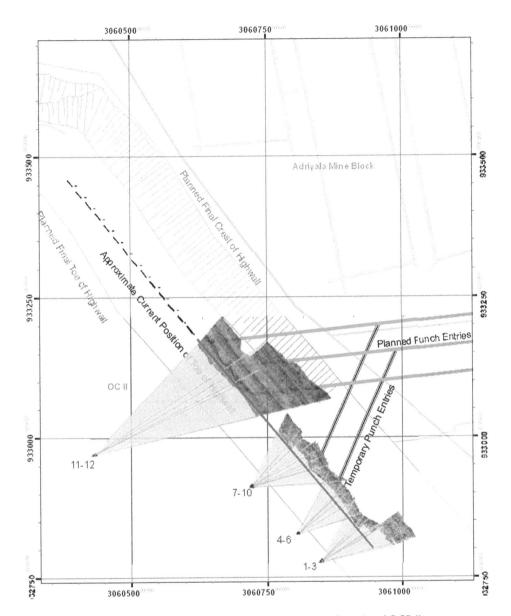

Figure 5.13 Approximate camera locations for the 12 stacks of OCP-II.

of the highwall resulted in areas being occluded in the images. This is demonstrated in Figure 5.14.

5.8.3 Image structural analysis

The structural analysis described in this section has been quoted from the works of Huddlestone-Holmes (2005) and Huddlestone-Holmes and Ketcham (2010). Analyses

Table 5.2 3D image resolution and accuracy for images collected at OCP-II.

Image stack	Approximate distance to toe of highwall (m)	Orientation error 50th percentile (°)		Spatial precision (mm)	Range resolution (mm)	Coverage (m)	Pixel size (mm)
		1m Joint	2m Joint				
I to 10	120	0.04	–#	3	9	37	11
11 to 12	300	0.21	0.04	8	24	94	27
11 to 12	400*	0.37	0.10	11	32	125	36

*Approximate distance to crest of highwall. #Too small to estimate reliably.

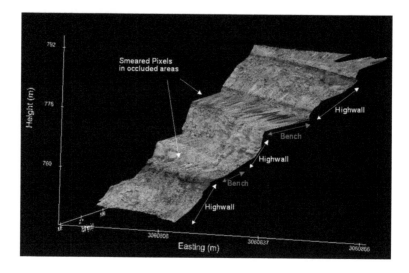

Figure 5.14 Image showing occluded areas on top of benches.

of the joint distribution were done on the image shown in Figure 5.5 of Section 5.4. This image was chosen as it covered a larger area when compared to other images, and the exposure was good with limited rubble. Joints were coloured, namely J1 in magenta, J2 in green and J3 in red. The joint statistics are shown in Table 5.3 and are plotted in Figure 5.15. No attempt had been made to measure cleat in the coal seams at this stage. J1 had very similar attributes to J1 in the CSIR-CIMFR (2005) report. However, the persistence and spacing of this joint set was greater in comparison to CSIR-CIMFR's study. J2 and J3 in this study were not represented in the CSIR-CIMFR (2005) report. When these two joint sets where combined, the dip direction of the combined set became 219°, which was similar to the dip direction of J3 in the CSIR-CIMFR (2005) report, though the dip was significantly shallower with greater spacing and persistence. The bedding was also measured and had a dip of 20° towards 52°, which matched with J2 in the CSIR-CIMFR (2005) report.

The distribution of the joints mapped in Figure 5.15 is not uniform. This is most likely a result of the actual distribution of joints in the rock mass. For example, it appeared that the coarser, cross-bedded, sandstone units had fewer joints than other

Table 5.3 Statistics for joints measured from image shown in Figure 5.5.

Set No.	No. Joints	Dip (°) Mean	Standard Deviation	Dip Direction (°)		Fisher K	R90	R99	Persistence (m)	Spacing (m)
				Mean	Standard Deviation					
J1	66	89.2	10.2	138.9	12.8	15.9	31.2	44.8	6.6 m	8.7
J2	32	61.1	14.3	199.3	11.8	17.1	30.1	43.0	4.5	3.9
J3	14	58.4	10.8	261.0	10.0	25.3	24.6	35.1	3.0	4.0

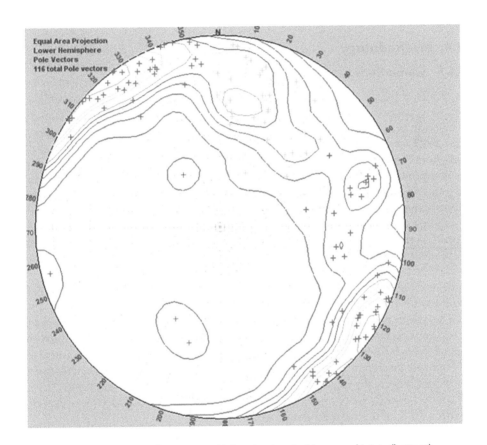

Figure 5.15 Image without mapped joints (top) and with mapped joints (bottom).

units. However, the observed distribution may be an artefact of the mapping process, at least in part. The coarse sandstone units appeared to weather quickly, rounding off edges; this might result in joints being less prominent than they would be in fresh rock. Parts of the slope were obscured by rubble, blasting damage and damage from mining equipment. Structures could not be mapped in these areas. The occluded areas in the images were the benches and were most likely covered in rubble and were not likely to have been useful for mapping structures, except that they made it difficult to trace structures that crossed the bench.

The joint density measured using Sirovision™ was lower than that presented in the CSIR-CIMFR (2005) report. There were several reasons why this might be the case: the CSIR-CIMFR (2005) report would have sampled a different part of the pit, which might have had a different joint population to that studied herein; the resolution of the image might have precluded less persistent joints from being mapped, reducing the number of observations; and the CSIR-CIMFR (2005) report might have sampled a more highly jointed area from what is a non-uniform joint distribution. Finally, when the number of structures sampled is small it can potentially affect the reliability of the statistics. As previously stated, the ISRM recommends that, in general, at least 200 structures per structure set should be sampled.

5.8.4 Methodology

5.8.4.1 Simulations replicating field measurements

The results of the structural mapping and analysis were imported into Siromodel. The joint set generator in Siromodel supports several models, but given the scarcity of information regarding spatial distribution of the joints, the Baecher model (Baecher et al., 1977) was used. This model assumes that the joints are distributed uniformly throughout the simulation volume with their total number adhering to a Poisson process (i.e. a binomial distribution tailored for large numbers of low probability events). The joints are assumed to exist as discs with log-normally distributed radii. This model is approximately valid, assuming that the distributions of the trace lengths themselves are log-normal. No confirmation of this has yet been performed for the structures mapped on the Ramagundam OCP-II highwall. Joint orientations were modelled using independent normal distributions for dip and dip direction. Note that Siromodel does not correct for any biases that are present in the data (e.g. orientation, censoring, length, etc.) although the inherent size bias associated with areal or line sampling is accounted for (assuming the Poisson disc model is valid).

Trials were performed to determine whether the simulated fracture networks produced simulated traces (Figure 5.16) on the simulated highwall that resembled those mapped with Sirovision™. Deviations between simulated and mapped traces typically indicate discrepancies between model assumptions and reality. The adjustment of parameters is normally required to achieve satisfactory statistical agreement. After several trials, the structure parameters in Table 5.4 were determined to achieve agreement to within 10% of the mapped parameters.

The bedding layer data was acquired from the CSIR-CIMFR (2005) report. Average values were used for depth of cover, thickness and orientation. The bedding structures were assumed to be infinitely persistent and the data are shown in Table 5.5.

Note that the friction angle and cohesion values used are conservative, as no information on these was available. A standard deviation of 10% has been applied to these values. Figure 5.17 shows an idealised model of the section of the highwall being simulated with nine benches of the wall. For the purposes of this demonstration, ten tunnel entries have been included in the modelling. These have been confined to seam IV. A Discrete Fracture Network (DFN) realisation is also shown.

A Monte Carlo simulation was conducted using 100 DFN realisations. As previously discussed, the polyhedral modelling algorithm utilised in Siromodel can detect

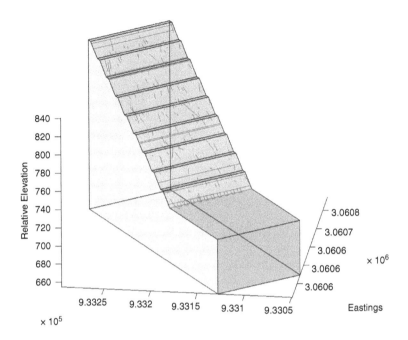

Figure 5.16 Simulated joint traces from a single DFN realisation.

Table 5.4 The joint set characteristics used in Siromodel.

Dip (°)	σ (Dip) (°)	DD (°)	σ (DD) (°)	Trace length (m)	No. of joints in simulation volume	Joint representation
89	10	139	13	8	2500	hexagon
61	14	199	12	5.5	2500	hexagon
58	11	261	10	3.6	2500	hexagon

Table 5.5 The bedding properties used in Siromodel (CSIR-CIMFR, 2005).

Seam	Depth of cover, m	Thickness m	Dip (°)	DD (°)	φ (°) (±10%)	C (kPa)
I	124.5	5.95	20	52	25	0
II	147	3.55	20	52	25	0
IIIA	192	1.35	20	52	25	0
III	219.5	11.6	20	52	25	0
IV	234.5	3.9	20	52	25	0

polyhedra, which form from the intersection of the polygons comprising the DFN. Only a few polyhedra (typically less than ten) were detected for each realisation.

Siromodel provides facilities to concatenate and perform several geotechnical analyses. Some are shown below in Figures 5.18, 5.19 and 5.20. In Figure 5.20, the vertical

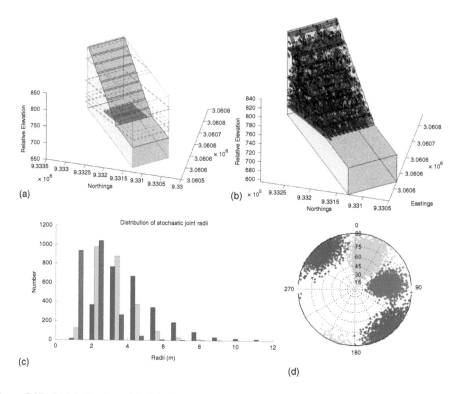

(a)

(b)

(c)

(d)

Figure 5.17 (a) Idealised model of the highwall with bedding layers and entry geometries; (b) Example realisation of a DFN used in this analysis; (c) Histogram of joint radii for DFN realisation in (b) and (d) Lower hemisphere stereonet for DFN realisation shown in (b).

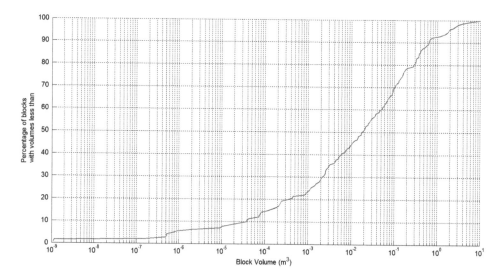

Figure 5.18 Cumulative block size distribution for kinematically free blocks.

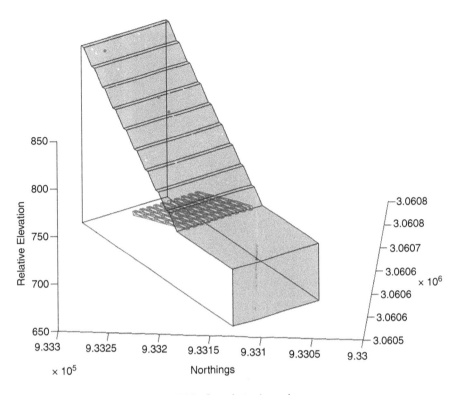

Figure 5.19 Cumulative hazard map.

axis shows the elevation above the lowest level in the model, so the bottom of the high-wall corresponds to an elevation of 60 m. The block size distribution curves represent the volumes of the polyhedra detected in each DFN realisation. The 90th percentile block volume is less than 10 m^3. The cumulative hazard map shows the locations and stability type of all kinematically free blocks detected for every DFN realisation, where red indicates type I (unstable), yellow type II (stable given friction) and green type III (kinematically stable). The hazard density map is particularly useful and represents the likelihood of hazards forming at a given location. The contours have been scaled such that the values can be interpreted as a percentage of the simulations predicting a hazard in a given location. Therefore, across the entire highwall, less than 3% of the simulations predicted hazard formation. No increased occurrence of hazards was detected in the regions of the entries.

This hazard density map is a proxy for 'probability of failure', but the reader is cautioned regarding this interpretation. Any numerical model is an approximation to reality and the fact that a failure is or is not observed in a simulation may not correlate in a predictable way with the reality and complexity of the heterogeneous rock mass. Therefore, the term percentage of simulations predicting failure is preferred by the authors.

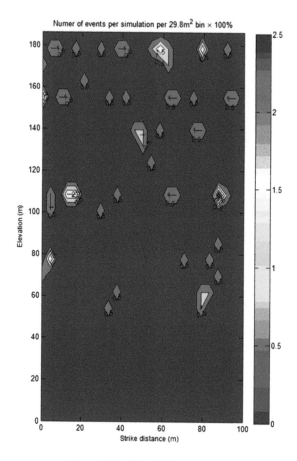

Figure 5.20 Hazard density map.

5.8.4.2   *Simulations assuming larger joints*

The preceding analysis assumes that the structural mapping has captured the salient characteristics of the discontinuities in the rock mass. As discussed in Section 5.2, there are a number of issues and biases that the practitioner needs to be aware of. Assessing the significance of these issues through techniques such as sensitivity analyses is extremely important for ensuring confidence in geotechnical analysis.

One of the more critical parameters associated with slope stability is structure size. A conservative analysis of the significance of this parameter has been conducted, assuming the trace lengths are much larger than mapped. Using the typical interburden thickness as a guide, the mean trace length of each of the three joint sets has been set to 25 m and the Monte Carlo simulation repeated. Figure 5.21 shows an example realisation where red blocks indicate type I (unstable), yellow type II (stable given friction) and green type III (kinematically stable) and Figure 5.22 shows the geotechnical analyses based on these realisations.

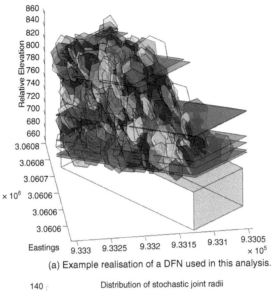

(a) Example realisation of a DFN used in this analysis.

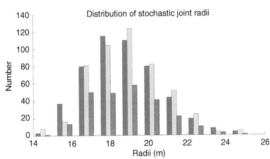

(b) Histogram of joint radii for DFN realisation in (a).

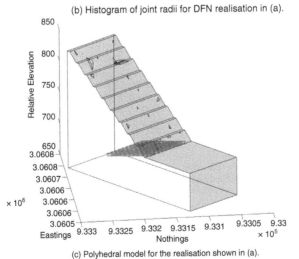

(c) Polyhedral model for the realisation shown in (a).

Figure 5.21   Realisation model of DFN.

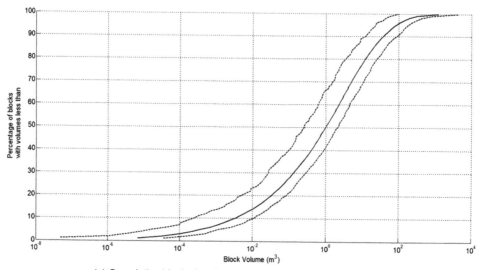

(a) Cumulative block size distribution for kinematically free blocks.

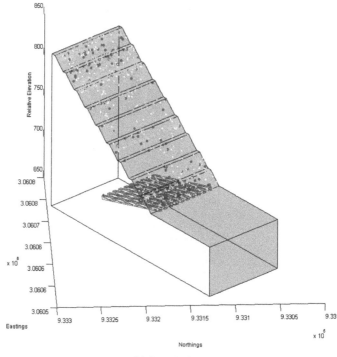

(b) Cumulative hazard map.

Figure 5.22 Several geotechnical analyses based on the Monte Carlo simulations for the more conservative scenario.

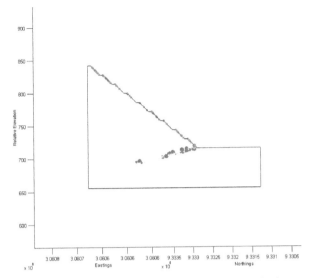

(c) Cumulative hazard map view along strike showing the increased
hazards associated with the entries.

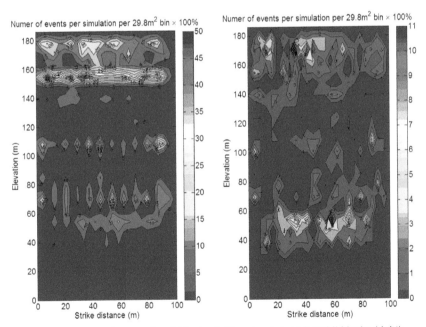

(d) Hazard density map for all blocks (left) and only type I and II blocks (right).
The vertical axis shows the elevation above the lowest level in the model, so
the bottom of the highwall corresponds to an elevation of 60 m.

Figure 5.22 Continued.

It is apparent that the increased structure size results in many more polyhedra forming. The 90th percentile block volume is 36 m³ but the 95% confidence interval for this parameter lies between 12 and 89 m³. The cumulative hazard map shows the locations and stability type of all kinematically free blocks detected for every DFN realisation, where red indicates type I (unstable), yellow type II (stable given friction) and green type III (kinematically stable). The hazard density map shows that across the entire highwall there are regions where up to 50% of simulations predict block formation but less than 11% of simulations predict type I and II blocks. This is particularly so for the region around seam II where the seam thickness is conducive to the formation of blocks but they are almost exclusively type III (i.e. kinematically stable) due to the bedding dipping into the wall. The structural modelling in this analysis has not discriminated between in-seam and interburden joint characteristics. This simplification is no doubt a contributing factor to these results.

Of particular note is Figure 5.22(c), which highlights the increased occurrence of hazards in the regions of the entries.

5.8.5 Simulations for interburden shale bedding

Another sensitivity analysis was conducted to determine the effects of shale bedding in the interburden above the seam IV entries of the OCP-II. For this Monte Carlo simulation, the original DFN parameters were used (i.e. joint traces as mapped in the field). One metre spacing was assumed for the shale bedding with orientation and friction angle identical to those assumed for the seam bedding.

Figure 5.23 shows the model geometry viewed along the strike, including the shale bedding directly above seam IV. Also shown are the (a) model geometry, (b) cumulative block size distribution curve (cumulative data) and (c) the hazard density maps for all blocks (left) and excluding type III blocks (right). Although the presence of the shale bedding has contributed to block formation relative to the analysis discussed in Section 5.8, less than 1% of simulations predict formation at any given location. There was also no significant incidence of type I or II block formation associated with the entries.

5.9 CONCLUDING REMARKS

The stability of the highwall is of the utmost importance to ensure safety during highwall mining operations. The scientific components discussed in this chapter concentrate on the guidelines for structural mapping and first pass stability analysis of slopes, depicted using the case example of the highwall at OCP-II of SCCL.

The first pass stability analysis presented in this chapter fits within a process for more sophisticated characterisation, as summarised in Figure 5.24. After the first pass analysis is complete, the practitioner should be armed with more detailed scenarios to investigate. More detailed Monte Carlo simulations (e.g. increased complexity in models, simulation volumes, etc.) may be practicable to allow the quantification of uncertainties with these predictions. By interrogating results based on polyhedral block size distributions, polyhedral stability analyses or other means, the identification

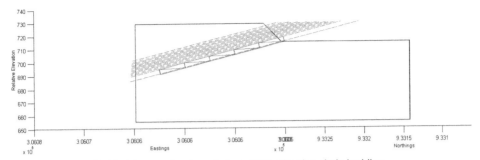

(a) Model geometry viewed along strike, showing shale bedding.

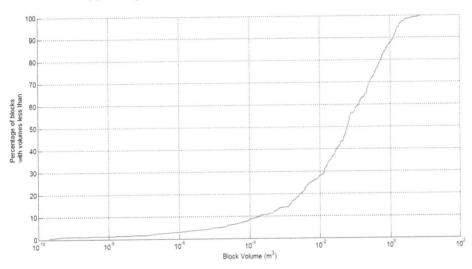

(b) Cumulative block size distribution curve.

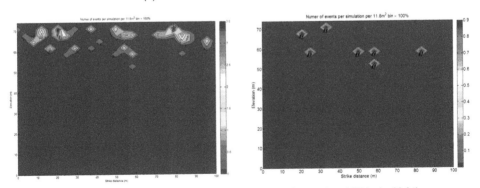

(c) Hazard density map for all blocks (left) and only type I and II blocks (right).

Figure 5.23 Model geometry viewed along strike above seam IV of OCP-II.

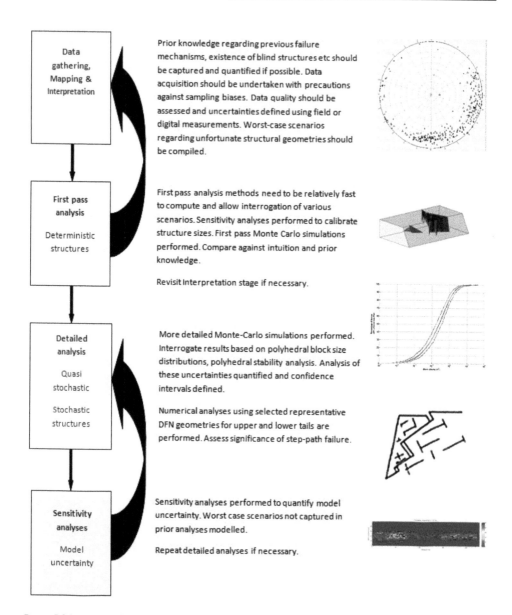

Figure 5.24 Process for integrating the structural mapping and analysis with more sophisticated slope stability studies (Elmouttie et al., 2012).

of DFN realisations associated with upper and lower tails of the predicted distribution, and therefore definition of confidence intervals, can take place. To capture the physics more accurately, numerical studies (e.g. stress-strain for fracture generation) of representative DFN realisations can take place.

This final processing stage can potentially be the most time-consuming. Model assumptions, including limitations in the physics, need to be explicitly enunciated and interrogated via sensitivity analyses. Worst case scenarios not previously captured need to be modelled for better predictability.

This could be further extended by conducting the following activities:

- More detailed structural mapping of discontinuities in all available images to better constrain joint statistics;
- Individually characterise the geotechnical domains, such as interburden versus seam joint characterisation;
- Explicit modelling of the cleat discontinuities;
- Sensitivity testing of joint surface properties (cohesion and friction angle);
- Sensitivity testing of joint spacing and persistence;
- Modelling of groundwater effects;
- Modelling of slope stability to account for *in situ* stress and rock mass failure.

Note: This chapter has been prepared based on chapter-3 of the AISRF Final Report: Agreement ST050173 (Baotang Shen, Shivakumar Karekal, Marc Elmouttie, John Loui Porathur and Pijush Pal Roy): Highwall mining design and development of norms for Indian conditions. *CSIRO Energy and CSIR-Central Institute of Mining & Fuel Research, EP156317/GAP/MT/DST/AJP/91/2011-12,* August 2015, 136 p.) wherein Dr. Marc Elmouttie, Research Team Leader, CSIRO Energy had contributed significantly. The authors are thankful to him for his kind consent to present his work on Sirovision™ and Discrete Fracture Network (DFN) modelling in this chapter for global readers.

Chapter 6

Australian design guidelines
and case studies

6.1 INTRODUCTION

This chapter describes the development of optimal design guidelines for highwall
mining and several case studies in Australia. This work started in the late 1990s and
it covers more than ten years of research conducted by the Commonwealth Scientific
and Industrial Research Organisation (CSIRO) with support from the Australian Coal
Association Research Programme (ACARP) and several major mining companies operating in Australia. The key contents of this chapter are extracted from a comprehensive
report by Duncan Fama et al. (2001). The aim of the study was to provide general
guidelines for highwall mining design for Australian mining conditions. The guidelines
will help guide mining engineers/mine managers through the process of feasibility evaluation to final execution of highwall mining in their mining lease. A number of new
tools for feasibility studies, including the Highwall Mining Index (HMI), the Auger
Mining Index and Percentage Extraction Estimate are presented in this chapter.

6.2 BRIEF HISTORY OF HIGHWALL MINING IN AUSTRALIA

Two types of highwall mining systems have been used. They are the Continuous Highwall Mining (CHM) System, which mines rectangular entries, usually about 3.5 m
wide, and the auger system, which creates individual or twin circular holes typically
1.5 to 1.8 metres in diameter. The current CHM systems originated in the USA in
the mid 1970s. They have been used for commercial mining since the early 1980s in
the USA. These technologies were first introduced into Australia in 1989, when BHP
Australia Coal Pty Ltd carried out a highwall mining trial at Moura Mine in Central
Queensland (Follington et al., 1994). Ten rectangular openings were driven into the
coal seam to depths of up to 30 m using an experimental continuous miner-based highwall mining system. Commercial highwall mining operations in Australia started two
years later in early 1991 when Callide Mine used an auger system to mine coal. Auger
mining was also employed at German Creek Mine and Oaky Creek Mine shortly after
the trial at Callide Mine. In early 1993, Oaky Creek Mine began operations with a
Joy CHM system (ADDCAR system) to mine coal on its leases.

    During the period 1991–2002, commercial highwall mining (CHM and auger)
was in operation in over 30 pits at 15 coal mines in Australia. Between 17 and 20

million tonnes have been produced from CHM, about 4.5 Mt from single augers and about 0.35 Mt from the twin auger. The mines employing highwall mining include:

- Oaky Creek (CHM and multi-level, endwall auger mining and single-pass trench auger mining)
- Moura (CHM and auger. Also using the Archveyor system, a new highwall mining system)
- German Creek (CHM and auger)
- Ulan (CHM)
- Callide (multi-level auger mining and twin auger mining)
- Collinsville (CHM)
- Yarrabee (CHM)
- Newlands (CHM)
- Warkworth (auger)
- Liddell (auger)
- Charbon (auger and CHM)
- Gregory (steep seam auger mining and twin auger mining)
- South Blackwater (CHM and endwall auger mining)
- Goonyella (endwall auger mining)
- Jellinbah (multi-level auger mining)

The CHM has an advantage in productivity and recovery rate. In Ulan Mine, a CHM system has routinely penetrated 500 m into the coal seam from the base of the highwall and has achieved a maximum production rate of 124,000 tonnes per month. The CHM, however, is very sensitive to ground conditions. In comparison, an auger system has exceeded 200 m penetration and produced up to 50,000 tonnes per month. Its lower penetration depth and production capacity is offset by its tolerance of difficult geological conditions. This makes it more suitable for highwall mining of a pit with unfavourable roof and floor conditions or for smaller reserves.

6.3 SCOPE OF THE CHAPTER

This chapter covers the following aspects of highwall mining:

- **Feasibility study.** The geotechnical evaluation to determine whether the site is suitable for highwall mining. A highwall mining index method is presented to help the decision-making process in Section 6.4.
- **Site investigation.** If the site is considered likely to be a suitable site for highwall mining, geological and geotechnical investigation in the mining reserve (e.g. drilling, highwall mapping, highwall hazard identification etc.) includes the gathering of data for layout design. The new data can also refine the results of the feasibility study. Guidelines for conducting the site investigation are outlined in Section 6.5.
- **Span stability assessment.** Evaluation of the stability of the entry spans, which are unsupported, to ensure the success of mining. This needs to be investigated before

the commencement of mining. Assessment tools and methodology are presented in Section 6.6.

- **Layout design.** A layout design to specify pillar size and whether barrier pillars are to be left between panels. An optimal layout design will maximise the recovery whilst minimising the risk of unacceptable instability. Pillar design tools and a new panel stability theory is presented to help in the evaluation in Section 6.7.
- **Design verification against past experience.** A number of panel/roof instabilities have been recorded in the past. A failure case has been back-analysed and the major causes of the instabilities summarised in Section 6.8. Lessons learned from this study may be used to provide a final check of the design for the current site.
- **Mining operation and feedback.** Monitoring of the stability of the highwall, spans and pillars during mining is critically important to a continuous, smooth mining operation. After mining operations have commenced with the designed layout, mining difficulties may occur if unexpected ground conditions are encountered. This may necessitate a change in layout design. Guidelines for conducting monitoring and information feedback during mining are provided in Section 6.9.

6.4 FEASIBILITY STUDY FOR HIGHWALL MINING

This section presents two practical tools for highwall mining feasibility study, i.e. the Highwall Mining Index (HMI) for CHM mining and the Auger Mining Index (AMI) for augering. The tools were designed to help mine managers and mining engineers in evaluating the feasibility of highwall mining in a given pit. Minimum conditions required for highwall mining are also discussed prior to the presentation of the tools.

Before employing highwall mining methods, two basic questions need to be answered:

- Is it *possible* for the pit to be mined using highwall mining methods?
- Is it *feasible* and *profitable* to mine the pit using highwall mining methods?

To answer the first question, one needs to examine the pit against the minimum conditions for highwall mining. To answer the second question, it is suggested that the highwall mining index for CHM or the auger mining index for augering be used to evaluate the overall condition of the pit. In general, economic issues are not the concern of this report, although many of the feasibility issues have economic implications. The following sections present the minimum conditions and the highwall mining index and auger mining index for the feasibility study.

6.4.1 Minimum conditions for highwall mining

There are some circumstances where highwall mining may not be feasible for safety, productivity, profitability or environmental reasons. Based on the experience of highwall mining in Australia over more than ten years, the following conditions are almost certain to preclude the use of highwall mining:

- **Poor roof condition.** Roof falls reduce mining progress, and can damage or bury mining equipment. If poor roof conditions are encountered, the site cannot be highwall mined using CHM. In general, auger mining is not sensitive to roof conditions. Even in extremely soft roof and floor conditions, leaving some coal in the roof and floor may facilitate successful mining operations. Conditions preventing highwall mining using CHM include: extremely weak to very weak rock, laminated rock, closely spaced rock jointing, major structures such as thrust faults close to the coal/roof interface, or extremely weathered rock.

- **Limited mining reserve.** If the mineable reserve is too small, then the cost of equipment mobilisation and pit preparation etc. is likely to be prohibitive. The minimum tonnage of the reserve depends on the availability of the mining equipment on site. As a general guide, for a new mine to use CHM highwall mining, a 250 m mineable pit length is considered to be the minimum. For a mine with existing highwall mining operations, a 100 m pit length is the minimum for CHM, and a 50 m pit length is the minimum for auger mining.

- **Seam methane content >10 m³/tonne.** As a general guide, a seam with methane content $>10\,m^3$/tonne is considered to be unsuitable for highwall mining if inertisation of the entries is not possible or economically feasible.

- **Seam dip >20°.** Existing CHM systems can mine coal seams dipping up to 15 degrees downward (BHP steep-dip machine at Moura Mine). The commonly used ADDCAR systems can mine up to 10 degrees downward for a penetration of 300 m, or up to 15 degrees for lesser penetration. The Archveyor system has been used to mine up to 18 degrees at Moura Mine. However, Nova Construction, Canada, has built a Highwall Miner that is designed to mine up to a 26-degree dip. Auger systems can mine down-dips of up to 20 degrees. A seam with a dip greater than 20 degrees cannot be mined using any existing highwall mining system in Australia without modification.

- **Unstable highwall and/or lowwall.** Highwall mining operations are controlled from the launch pad immediately beside the highwall. A grossly unstable highwall, or a large-scale lowwall instability in combination with a narrow pit, will preclude highwall mining operations in that part of the pit.

- **Seam thickness <1.0 m.** A seam less than 1.5 m thick cannot be mined using CHM, and seams less than 1.0 m cannot be mined economically using an auger.

- **Pit width <30 m.** A minimum pit width of 30 m is required to mobilise and operate the highwall mining system.

- **Stone bands in coal seam.** Hard stone bands thicker than 0.2 m in the coal seam may prevent the use of an auger system. In addition, seams containing dirt bands or clay seams sensitive to moisture occurring anywhere across the hole diameter are difficult to mine using an auger. CHM mining can tolerate stone bands although incendiary sparking may be an issue.

If any of the above conditions are encountered at the given site, the site is likely to be rejected for further consideration for highwall mining. If, however, none of these adverse conditions are present, a second evaluation should be carried out to quantify the suitability of the site for highwall mining and the likelihood of success. The highwall mining index (HMI) for CHM and the Auger Mining Index (AMI), discussed below, provide tools for this purpose.

6.4.2 Highwall mining index (HMI) for CHM

The suitability of a site for highwall mining depends on a number of geological, geotechnical, operational and commercial parameters. In particular, the parameters listed above must individually satisfy the minimum requirement for highwall mining, but they may have different importance in the overall assessment. For this reason, all relevant parameters need to be assessed systematically, since some combinations of parameters may still threaten a safe and profitable mining operation. A practical way of doing a systematic assessment is to quantify the condition of each parameter and project the scores of all the parameters into one index value. This index value will reflect the overall condition of the site for highwall mining, and can help in deciding whether to employ highwall mining at this site. The HMI method was proposed by Seib (1992). It has been modified for this report based on the experience of the last nine years of highwall mining practice in Australia.

The HMI focuses on CHM operations, which are the most widely used in the industry. Where geotechnical conditions do not permit the application of CHM, an auger system may be considered. It is emphasised that the HMI system is intended as the second step of the initial evaluation, not as the definitive assessment. It is also emphasised that there will always be uncertainty arising from unexpected geological structures or other variability in conditions which may be encountered during mining operations.

The user of this method requires an adequate understanding of all of the parameters listed in the next section.

Key parameters

The 15 key parameters, which have had a major effect on the success of highwall mining, including the eight parameters listed in Section 6.4.1, are summarised and grouped below in four categories based on their impact on decision-making. Guidelines for the assignment of numerical values to these parameters are given in Table 6.1.

Group A – Most critical parameter. The roof condition is the single most critical parameter for highwall mining using CHM. Many of the unsuccessful highwall mining operations in the past were a direct or indirect result of poor roof conditions.

> **A1 Roof condition** – Roof falls can cause frequent withdrawal of the mining system and shut down the mining process. In the worst case, they can damage or bury the CHM system. Roof falls are often the major cause of limited penetration. Experience has shown that good roof is one of the most important parameters for successful highwall mining.

Group B – Critical parameters. These parameters are considered critical in deciding whether highwall mining operations will be possible in a given pit.

> **B1 Mineable reserve** – If the reserve is small, the fixed costs such as pit preparation and mobilisation will be a much larger proportion of the total cost and therefore mining may not be economically feasible.
>
> **B2 Seam gas** – Seam gas directly affects the safety of highwall mining operations, as for underground mining. If the seam has a very high methane content, the seam

simply cannot be mined using current highwall mining systems as it is impossible to dilute and ventilate the methane adequately.

B3 **Seam dip** – If the seam dip is greater than 15 degrees downward, the seam cannot be mined with systems currently in use in Australia. Gently dipping coal seams may be mined. Very limited experience of mining upward inclined seams is available. However, an upward slope greater than 5 degrees is considered to be unsuitable for highwall mining.

B4 **Highwall/lowwall stability** – Since highwall mining systems operate at the foot of the highwall and the lowwall, an unstable highwall or lowwall is a serious threat to safety. No operations will be viable under such conditions. For small-scale rock falls from the highwall or lowwall, protective measures need to be implemented.

Group C – Sub-critical parameters. Parameters in this group significantly affect productivity and profitability. Although they are not significant enough to preclude employing highwall mining, they often dominate the economic feasibility of using this mining method.

C1 **Seam continuity** – Faulting that displaces the seam by more than 1 m often stops highwall mining, and hence significantly limits the penetration distance and resource recovery.

C2 **Pit condition** – Highwall mining requires adequate space and a clean floor in the pit for equipment movement and storage. If the pit is too narrow, or the floor is too soft and muddy, excavation of the lowwall or digging into the floor may be required. This may not be geotechnically or economically feasible.

C3 **Seam quality** – This parameter will directly affect the profitability of highwall mining. Depending on the market price, a low-quality coal seam may provide little profit margin. Low quality could be a result of low coal rank and/or dirt bands in the coal seam, which dilute the overall coal quality. Highwall mining can improve product quality by selectively mining the higher quality section of the seam.

C4 **Floor condition** – Soft floor conditions cause the continuous miner to trip and hence slow or even stop mining. They can also undermine pillar/panel stability.

C5 **Groundwater** – Excessive water flow into a mining entry may force the withdrawal of the system and termination of the entry.

Group D – Important parameters. Parameters in this group are non-critical parameters to decision-making but they can have a substantial impact on the productivity and profitability of the operation.

D1 **Coal strength** – This parameter will affect the resource recovery since it affects the pillar size.

D2 **Cover depth** – This parameter affects the pillar size required to maintain the panel stability.

D3 **Subsidence** – In the long term, subsidence may occur in many highwall mined pits. If long-term stability is required, measures such as back-filling or leaving large pillars may be necessary. Back-filling entries will increase operating costs, whereas using oversized pillars will reduce the resource recovery rate.

D4 Seam thickness – The mineable seam thickness for single-pass CHM ranges from 1.5 m to 4 m. Optimal seam thickness helps to achieve the best productivity and resource recovery rate. It is also desirable that seam thickness is uniform within a reserve.

D5 Mobilisation and service – Local or onsite availability of the mining system and service could affect the mining productivity and the operating cost.

6.4.3 Conditions of the key parameters

Each of the 15 key parameters is categorised into four conditions for a given pit: Excellent, Good, Fair and Poor. The HMI user assigns a rating in the range of 0–100 to each parameter based on its condition:

Excellent:	80–100,	default = 90
Good:	60–80,	default = 70
Fair:	40–60,	default = 50
Poor:	0–40,	default = 20

Where there is insufficient information to make a judgement a default value is given to each parameter. A description of the different conditions for the 15 key parameters is given in Table 6.1.

6.4.4 Calculating the Highwall Mining Index (HMI)

The HMI is a mean value of the rating of each of the 15 parameters. The 15 parameters have been weighted according to their relative importance in the decision-making process. The individual weight factor of each parameter has been assigned according to the rank of the group that it belongs to. Parameters in the same group have been given the same weight.

In practice, if a critical parameter is very poor (e.g. if the roof condition is extremely bad), then it will probably not be wise to proceed with highwall mining, even though all other parameters are excellent. The HMI reflects such extreme cases by using a weighted geometric mean value of the 15 parameters, rather than a weighted average value. The geometric mean value of the parameters will approach zero if one of the parameters is close to zero, whereas the average value would still be high even when the value of one of the parameters is close to zero.

The geometric mean value of the parameters in each of the four groups is given by:

$$\overline{A} = A$$
$$\overline{B} = (B1 \cdot B2 \cdot B3 \cdot B4)^{1/4}$$
$$\overline{C} = (C1 \cdot C2 \cdot C3 \cdot C4 \cdot C5)^{1/5}$$
$$\overline{D} = (D1 \cdot D2 \cdot D3 \cdot D4 \cdot D5)^{1/5}$$

In these expressions, each parameter in each group is given the same weight.

The HMI is the weighted mean value of the four groups:

$$HMI = (\overline{A}^{1.6} \cdot \overline{B}^{1.2} \cdot \overline{C}^{0.8} \cdot \overline{D}^{0.4})^{1/4}$$

Table 6.1 Rating guidelines for the 15 key parameters included in the highwall mining index.

Rank	Parameter	Condition				
		Excellent (100 ≥ rating > 80)	Good (80 ≥ rating > 60)	Fair (60 ≥ rating > 40)	Poor (40 ≥ rating > 0)	No-go (rating = 0)
A	Roof condition	Strong to very strong roof rocks; thickly bedded; joints with wide spacing. No alternation and weathering.	Moderately strong roof rock; medium bedded; joints with moderate spacing. Slight alternation and weathering.	Moderately weak roof rocks; thinly bedded; joints with close spacing. Alternation and weathering.	Weak roof rocks; very thinly bedded; joints with very close spacing. Strong alternation and weathering.	Extremely weak to very weak rock; laminated; joints with very close spacing. Severe weathering.
B1	Mineable reserves	>1,500 m highwall	500 m to 1,500 m highwall	250 m to 500 m highwall	100 m to 250 m highwall	<100 m highwall
B2	Seam gas (methane)	Very low methane content (<0.1 m³/t). No concern for mining.	Low methane content in general (0.1–1 m³/t). Higher gas only in localised pockets.	Seam methane = 1–5 m³/t. Gas problems could be encountered occasionally in reserve.	Seam gas (5–10 m³/t). Gassy seam and gas problems are expected to be encountered throughout.	Seam methane >10 m³/t
B3	Seam dip	+0° to −4°	+0° to +1°, or −4° to −7°	+1° to +2°, or −7° to −10°	+2° to +4°, or −10° to −15°	>+4°, <−15°
B4	Highwall/lowwall stability	Smooth highwall. No hanging blocks. No major joint sets or faults. No risk of large-scale lowwall instability.	Localised loose rock blocks. Slight weathering. Minor joints/cracks. No discernible risk of lowwall instability.	Common small rock falls. Dense jointing. No risk of large-scale highwall instability. Very low risk of lowwall instability. Possible block falls.	Large rock falls. Weathered. Wedges and slabs formed by joints. Some risk of local lowwall instability but manageable.	Grossly unstable highwall. Significant risk of large-scale lowwall instability.
C1	Seam continuity	Seam is consistent with no faulting or folds.	Structures which slightly reduce seam thickness. Only small faults and undulations.	Structures which reduce seam thickness by up to 50%. Seam changes direction by 5° in folds.	Structures which reduce thickness over 50%. Faults and distinct folds are common.	Seam heavily affected by faulting and folding.
C2	Pit condition, preparation	No loose material, dry, no floor work.	<0.5 m loose material. Water puddles. Shave floor.	<2 m loose material. Pumping required. Some mud. Soft floor.	>2 m loose material. Major pumping and mud clean up. Floor repairs required.	Pit width <30 m.
C3	Seam quality	Medium to hard export coking coal. No in-seam parting or stone bands.	Soft coking coal and export steaming coal. One or two stone bands.	Any unwashed product which is saleable. Many partings and stone bands.	Coal not in previous categories which must be washed for sale. Many stone bands.	

C4	Floor condition	Strong floor rocks. Unaffected by water.	Weak to moderately strong floor rocks. May be wet but not muddy.	Weak floor rocks. Occasionally muddy.	Very weak floor rocks (clay). Wet and muddy throughout.	
C5	Groundwater	Dry environment. No water gets into the mining entries.	Seasonal dry and wet environment. Some water into entries but is manageable.	Generally wet environment. Some water problems could be expected at fault zones.	Very wet environment. Water gets into entries from the pit and from the ground. It may flood the entry.	
D1	Coal strength	Lab coal uniaxial compressive strength (UCS) >30 MPa. Limited coal joints with wide spacing.	Lab coal UCS = 20–30 MPa. Coal joints with moderate spacing.	Lab coal UCS = 10–20 MPa. Coal joints with close spacing.	Lab coal UCS <10 MPa. Heavily cleated. Coal joints with very close spacing.	
D2	Cover depth (overburden + spoil)	Overburden depth <80 m in the reserve.	Overburden depth 50–100 m.	Overburden depth 60–120 m, possible stress concentration near highwall.	Overburden depth 80–150 m, possible significant stress concentration near highwall.	
D3	Subsidence	Surface subsidence not an issue.	Minor subsidence effects but all contained on lease. Minimal rehabilitation efforts.	Some subsidence or water effects off site but controllable. Rehabilitation could be costly.	No subsidence is allowed. Significant water effects off site. Major rehabilitation costs. Subsidence control is critical.	
D4	Seam thickness	>3.5 m	2.5 m to 3.5 m	2.0 m to 2.5 m	1.5 m to 2.0 m	<1.5 m
D5	Mobilisation and service	Equipment, service capacity and skilled mining crew are available on site.	Equipment, service capacity and skilled mining crew are available in mines nearby.	Equipment, service capacity and skilled mining crew are available interstate.	Equipment and service capacity are available overseas. New mining crews need to be trained.	

*Terminology follows the International Society for Rock Mechanics (ISRM) standard.

The weighting assigned to each group is given by the exponential factors of 1.6, 1.2, 0.8 and 0.4 to groups A, B, C and D respectively.

The value of HMI ranges between 0 and 100. It represents the attractiveness of a site for highwall mining with CHM. A score of 60 or greater indicates a prospect with a high potential for highwall mining, worthy of a detailed feasibility study. A score of less than 40 indicates a poor site, which would be unlikely to produce coal safely and profitably using the current methods of highwall mining using CHM.

A score between 40 and 60 indicates that the site is not very attractive from a purely technical point of view. However, other non-technical issues, such as an urgent need for coal, or an increase in market price, may weight this site towards a potential mining site.

6.4.4.1 Confidence rating for the Highwall Mining Index

Assigning a rating (0–100) of the 15 key parameters in the HMI calculation is a subjective process. It is affected by the information available to the person who makes the judgement. Some parameters can be rated easily with full confidence whereas others can only be rated with a limited level of confidence. For instance, "pit condition" can be rated easily with a confidence level of near 100% since the pit and its floor are likely to be fully exposed and can be inspected visually. "Seam continuity", however, may only be rated with a low confidence level as the understanding of local geology may not be sufficient, depending on the amount of information available.

New mines are likely to have a smaller data set and hence the confidence level in the values assigned to many parameters may be low. By contrast, proposed operations at an existing mining site may have detailed information available and so can assign higher confidence levels to their data.

Analysing the individual contributions to the confidence level of HMI will help to identify the parameters in need of further assessment.

To calculate the confidence level of HMI, each parameter is given a confidence level on a scale from 0 to 100%. The confidence level of an individual parameter is weighted in the same way as it is weighted in the HMI calculation. The total confidence level of the HMI is calculated by the weighted geometric mean of the confidence levels of each of the 15 parameters.

$$C_{\overline{A}} = C_A$$
$$C_{\overline{B}} = (C_{B1} \cdot C_{B2} \cdot C_{B3} \cdot C_{B4})^{1/4}$$
$$C_{\overline{C}} = (C_{C1} \cdot C_{C2} \cdot C_{C3} \cdot C_{C4} \cdot C_{C5})^{1/5}$$
$$C_{\overline{D}} = (C_{D1} \cdot C_{D2} \cdot C_{D3} \cdot C_{D4} \cdot C_{D5})^{1/5}$$

where C_X is the confidence level of parameter X.

The overall confidence level of HMI is given by:

$$C_{HMI} = (C_{\overline{A}}^{1.6} \cdot C_{\overline{B}}^{1.2} \cdot C_{\overline{C}}^{0.8} \cdot C_{\overline{D}}^{0.4})^{1/4}$$

A confidence score greater than 60% indicates that, in the opinion of the person calculating the HMI, the calculated HMI is likely to be reasonably reliable. A score of less than 50% indicates a level of uncertainty in the evaluated HMI, which needs

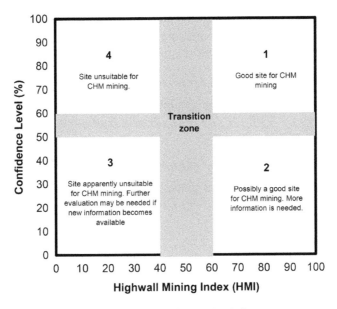

Figure 6.1 HMI – confidence level chart.

to be resolved by getting additional information to better define the rating of those parameters for which there is currently greatest uncertainty.

6.4.4.2 Decision-making based on the HMI and its confidence value

For a given site, the calculated HMI and its confidence level are plotted as a point in the HMI confidence level chart as shown in Figure 6.1. The chart area is divided into four regions and a transition zone. A subjective assessment is provided, depending on the region where the plotted point falls, as shown in Figure 6.1.

Region 1 – High HMI with high confidence level

Based on the available information, the pit evaluated is a good potential site for highwall mining. Detailed mine planning and financial analysis should be carried out.

Region 2 – High HMI with low confidence level

Based on the available information, the pit may be a good potential site for highwall mining, but significant uncertainty exists in the judgement rating of many parameters. The low confidence parameters need to be identified and carefully investigated. Some preliminary site investigation may be needed to reduce the uncertainty in these parameters.

Region 3 – Low HMI with low confidence level

Based on the available information, the pit appears to have low potential for highwall mining. However, the low confidence in the parameter rating means that this assessment could change as more information becomes available.

Region 4 – Low HMI with high confidence level

Based on the available information, the pit is not suitable for highwall mining. No further consideration of this pit for CHM mining is necessary. However, it may be suitable for auger mining.

Transition zone – Medium HMI with medium confidence level

Depending on where an assessment plots, a decision should be made as to whether to proceed to obtain more information. Other non-technical issues may also require consideration here.

6.4.4.3 Example case

A pit in Central Queensland was to be considered for highwall mining using CHM. Based on previous site investigations in neighbouring regions, as well as observation of the current pit, the following information was available to the mining engineer:

Roof condition: Based on the observation of the highwall face, the immediate roof is thinly bedded weak mudstone, which has a thickness of about 0.5 m. It is overlain by a layer of stronger siltstone. The main roof overlying the siltstone is a thick layer of more competent, intact sandstone and interbedded siltstone. The immediate mudstone roof layer may create some roof falls, but the roof falls are not expected to extend through to the siltstone or sandstone layer above. Some joints exist in the roof but are not intense. **Rating:** 70 (good), **confidence level:** 70%.

Mineable reserve: The total length of the reserve is about 750 m along the highwall. There is no limitation to mining penetration, which could extend beyond 300 m. There are no other constraints, such as major faults. **Rating:** 80 (good), **confidence level:** 95%.

Seam gas: The seam has a low methane content. Previous highwall mining of the same seam in nearby areas did not experience major problems with seam gas. **Rating:** 75 (good), **confidence level:** 80%.

Seam dip: Based on information in the neighbouring areas, the seam dip is expected to be in the range of 4.5° to 8.0° down dip. **Rating:** 75 (good), **confidence level:** 70%.

Highwall stability: The final pit highwall was created using pre-split blasting techniques. The highwall face is generally smooth throughout the pit. No major faults exist that have the potential to create major highwall instability. Two joint sets were observed in some areas, which may create potential localised face instability. There are also some detached blocks from the highwall face and slumping of material from the highwall crest. There is no risk of lowwall failure. **Rating:** 75 (good), **confidence level:** 70%.

Seam continuity: Based on the observation of the final highwall and the previous strips, no major faults and folds are expected in the mining reserve. No in-seam drilling has been done. **Rating:** 90 (excellent), **confidence level:** 60%.

Pit condition: There is sufficient pit space (>60 m) between the lowwall and highwall for highwall mining operations. The pit floor is competent with no soft material in general. Some rain water and mud has accumulated near the highwall toe. Pumping of the water is required, but this should be fairly easy using equipment already on site. **Rating:** 100 (excellent), **confidence level:** 100%.

Coal quality: The seam is good-quality, low ash coking coal. No stone bands or partings were observed. **Rating:** 90 (excellent), **confidence level:** 95%.

Floor condition: The immediate floor is composed of carbonaceous mudstone up to 0.2 m thick, with a sheared clay layer within 0.5 m of the seam floor. The floor material is generally weak. The seam and the floor are likely to be wet. **Rating:** 60 (fair), **confidence level:** 50%.

Groundwater: The mine site is generally dry and groundwater is not a major concern. However, during the wet season, surface water may accumulate quickly and find its way to the mining entry from the pit floor or through geological structures. **Rating:** 70 (good), **confidence level:** 70%.

Coal strength: Extensive studies were conducted on coal strength of the seam in nearby areas. The seam is considered to be medium strong (UCS = 20–30 MPa). The seam in this pit is well cleated and has regular joints. **Rating:** 75 (good), **confidence level:** 90%.

Cover depth: The cover depth is about 55 m at the highwall and is estimated to be around 90 m at the maximum penetration of 300 m. There is no spoil pile on the surface and the highwall is straight without any benches. No abnormal stress concentration is expected near the highwall. **Rating:** 80 (good), **confidence level:** 90%.

Subsidence: There is no concern about future subsidence of the mined ground. There is no water reserve that could be affected by any possible subsidence. **Rating:** 100 (excellent), **confidence level:** 95%.

Seam thickness: The seam at the highwall exposure has a thickness of 3.2–3.4 m. The thickness is not expected to change significantly in the reserve. **Rating:** 80 (good), **confidence level:** 60%.

Mobilisation and service: The equipment, skilled mining crews and service capacity are on site. **Rating:** 90 (excellent), **confidence level:** 90%.

Using the above information, the process of calculating HMI is demonstrated below.

The average rating of the four groups:

$$\overline{A} = A = 70$$
$$\overline{B} = (B1 \cdot B2 \cdot B3 \cdot B4)^{1/4} = (80 \cdot 75 \cdot 75 \cdot 75)^{1/4} = 76$$
$$\overline{C} = (C1 \cdot C2 \cdot C3 \cdot C4 \cdot C5)^{1/5} = (90 \cdot 100 \cdot 90 \cdot 60 \cdot 70)^{1/5} = 81$$
$$\overline{D} = (D1 \cdot D2 \cdot D3 \cdot D4 \cdot D5)^{1/5} = (75 \cdot 80 \cdot 100 \cdot 80 \cdot 90)^{1/5} = 84$$

The highwall mining index (HMI):

$$HMI = (\overline{A}^{1.6} \cdot \overline{B}^{1.2} \cdot \overline{C}^{0.8} \cdot \overline{D}^{0.4})^{1/4} = (70^{1.6} \cdot 76^{1.2} \cdot 81^{0.8} \cdot 84^{0.4})^{1/4} = 75$$

Table 6.2   Rating of the 15 key parameters and their confidence level, example pit 1.

Rank	Parameter	Judgement rating	Confidence level (%)
A	Roof condition	70	70
B1	Mineable reserve	80	70
B2	Seam gas	75	80
B3	Seam dip	75	70
B4	Highwall stability	75	70
C1	Seam continuity	90	60
C2	Pit condition (pit preparation)	100	100
C3	Coal quality	90	95
C4	Floor condition	60	50
C5	Groundwater	70	70
D1	Coal strength	75	90
D2	Cover depth	80	90
D3	Environmental aspect – subsidence	100	95
D4	Seam thickness	80	70
D5	Mobilisation and services	90	90

Highwall Mining Index (HMI) = 75
Confidence level of HMI = 73%

The confidence level of each group:

$$C_{\overline{A}} = C_A = 0.7$$
$$C_{\overline{B}} = (C_{B1} \cdot C_{B2} \cdot C_{B3} \cdot C_{B4})^{1/4} = (0.7 \cdot 0.8 \cdot 0.7 \cdot 0.7)^{1/4} = 0.72$$
$$C_{\overline{C}} = (C_{C1} \cdot C_{C2} \cdot C_{C3} \cdot C_{C4} \cdot C_{C5})^{1/5} = (0.6 \cdot 1.0 \cdot 0.95 \cdot 0.5 \cdot 0.7)^{1/5} = 0.72$$
$$C_{\overline{D}} = (C_{D1} \cdot C_{D2} \cdot C_{D3} \cdot C_{D4} \cdot C_{D5})^{1/5} = (0.7 \cdot 0.9 \cdot 0.9 \cdot 0.95 \cdot 0.7 \cdot 0.9) = 0.86$$

The confidence level of HMI:

$$C_{HMI} = (C_{\overline{A}}^{1.6} \cdot C_{\overline{B}}^{1.2} \cdot C_{\overline{C}}^{0.8} \cdot C_{\overline{D}}^{0.4})^{1/4} = (0.7^{1.6} \cdot 0.72^{1.2} \cdot 0.72^{0.8} \cdot 0.86^{0.4})^{1/4}$$
$$= 0.73 = 73\%$$

Table 6.1 and Table 6.2 summarise the individual rating of each of the 15 parameters and the resultant HMI and its confidence level.

The results indicate that, for the pit evaluated, the HMI is 75 and the confidence level is 73%. When plotted on the HMI confidence chart, this pit falls into the centre of Region 1 (Figure 6.2). Hence, the pit is considered to be a good potential site for highwall mining. However, some parameters with low confidence levels may need to be examined in more detail before final mine planning and financial analysis is undertaken. These parameters include roof condition, seam continuity, floor condition, and seam thickness. Geotechnical investigations to better evaluate these parameters would be desirable before a final commitment to highwall mining.

Clearly, the sensitivity of the HMI to any of the parameters can be tested by changing one or more parameters and recalculating the HMI.

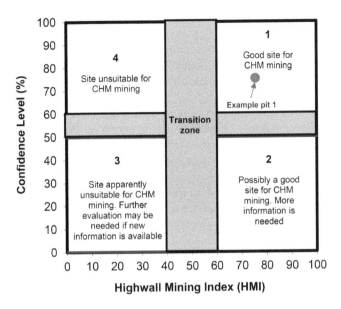

Figure 6.2 Highwall mining index of example site 1.

A highwall mining operation was eventually carried out in the pit, after a thorough site investigation and geotechnical design. The pit was mined successfully.

6.4.5   Evaluation of auger mining

Auger mining is more tolerant of adverse geotechnical conditions than CHM mining. For instance, roof conditions are generally not an issue for auger mining as it can tolerate almost all roof conditions. This is because an auger creates circular holes which are inherently more stable than the rectangular entries created by CHM and do not expose an unsupported span in the conventional sense.

The traditional surface auger mining requires some minimum conditions as outlined in Section 6.4.1, such as seam gas, pit width, highwall/lowwall stability, seam continuity etc.

A disadvantage of auger mining is its limited penetration distance (<200 m) and relatively low production capacity compared with CHM. However, due to its tolerance of bad roof and floor conditions, auger mining typically achieves the designed penetration distance more consistently. This is in comparison with often highly variable penetration distance when using CHM. The disadvantage of the short penetration distance can be compensated for to some degree by the consistency.

6.4.5.1   Auger Mining Index (AMI)

To help in a preliminary feasibility study for auger mining, an Auger Mining Index (AMI) is proposed in this section. The AMI is mainly based on the HMI, but some parameters in the HMI are re-evaluated and/or given a new rating based on their effect on the auger mining.

The parameters in Table 6.1 that are no longer important or relevant to auger mining include:

Roof condition – Auger mining tolerates almost all types of roof conditions. In general, auger mining is not sensitive to roof conditions. Even in extremely soft roof and floor conditions, leaving some coal in the roof and floor ensures successful mining operations.

Floor condition – Similarly, auger mining is not sensitive to the floor conditions.

Subsidence – Due to the limited penetration, the extent of any subsidence of an auger mining panel will be limited to the vicinity of the highwall. Hence, its impact on the environment may not be a major concern.

Mobilisation and service – An auger system is smaller and more mobile than a CHM system. In addition, most Australian mines have skilled crew to operate and service the system. Mobilisation and service is therefore not a major concern for auger mining.

These four parameters are therefore deleted from the key parameter list for AMI. The parameters in Table 6.3, which need to be re-rated for auger mining, include:

Mineable reserve – The required mineable reserve (length of highwall) for auger mining should be less than that required by CHM, mainly due to the relatively low cost of the mobilisation of an auger system and the fact that an auger can go through difficult ground conditions and achieve consistent penetration depths.

Cover depth – Within a penetration depth of 200 m, the variation of the cover depth will be less than that for CHM. Therefore, the range of the cover depths for each condition (i.e. excellent, good, fair and poor) is redefined.

Seam thickness – The optimal working seam thickness for auger mining using the currently available equipment is 1.5–2.5 m. Seams less than 1.0 m are not mineable. Seams between 3.5 m and 5.0 m may be mined by two-pass auger mining without difficulty, but a CHM may be more suitable for this seam thickness. Seams greater than 5 m may require more complex layout beyond two passes and can cause operational and geotechnical concerns.

A new parameter is added into the AMI. It is:

Stone bands – Stone bands in the coal seam have a significant effect on auger mining performance. Very weak stone bands in the seam are likely to cause instability of the auger holes. Seams containing hard stone bands thicker than 0.2 m are difficult to cut using an auger.

The Auger Mining Index (AMI) uses 12 parameters, which are divided into three groups, as follows:

Group A – Critical parameters
  A1 Mineable reserve
  A2 Seam gas
  A3 Seam dip
  A4 Highwall/lowwall stability

Table 6.3 Rating guidelines for the 12 key parameters included in the auger mining index.

Rank	Parameter	Condition				
		Excellent (100 ≥ rating > 80)	Good (80 ≥ rating > 60)	Fair (60 ≥ rating > 40)	Poor (40 ≥ rating > 0)	No-go (rating = 0)
A1	Mineable reserves	>1,000 m highwall	250 m to 1,000 m highwall	100 m to 250 m highwall	50 m to 100 m highwall	<50 m highwall
A2	Seam gas (methane)	Very low methane content (<0.1 m³/t). No concern for mining.	Low methane content in general. (0.1–1 m³/t). Higher gas only in localised pockets.	Seam methane = 1–5 m³/t. Gas problems could be encountered occasionally in reserve.	Seam gas (5–10 m³/t). Gassy seam and gas problems are expected to be encountered throughout. May be necessary to inertinise the hole in advance, but at cost.	Seam methane >10 m³/t. Inertinisation is too costly or impossible.
A3	Seam dip	+0° to −4°	+0° to +1°, or −4° to −7°	+1° to +2°, or −7° to −10°	+2° to +4°, or −10° to −20°	>+4°, <−20°
A4	Highwall/lowwall stability	Smooth highwall. No hanging blocks. No major joint sets or faults. No risk of large-scale lowwall instability.	Localised loose rock blocks. Slight weathering. Minor joints/cracks. No discernible risk of lowwall instability.	Common small rock falls. Dense jointing. No risk of large-scale highwall instability. Very low risk of lowwall instability. Possible block falls.	Large rock falls. Weathered. Wedges and slabs formed by joints. Some risk of local lowwall instability but manageable.	Grossly unstable highwall. Significant risk of large-scale lowwall instability.
B1	Seam continuity	Seam is consistent with no faulting or folds.	Structures which slightly reduce seam thickness. Only small faults and undulations.	Structures which reduce thickness by up to 50%. Seam changes direction by 5° in folds.	Structures which reduce thickness over 50%. Faults and distinct folds are common.	Seam heavily affected by faulting and folding.
B2	Pit condition, preparation	No loose material, dry, no floor work.	<0.5 m loose material. Water puddles. Shave floor.	<2 m loose material. Pumping required. Some mud. Soft floors.	>2 m loose material. Major pumping and mud clean up. Floor repairs.	Pit width <30 m.
B3	Seam quality	Medium to hard export coking coal. No in-seam parting or stone bands.	Soft coking coal and export steaming coal. One or two stone bands.	Any unwashed product which is saleable. Many partings and stone bands.	Coal not in previous categories which must be washed for sale. Many stone bands.	
B4	Groundwater	Dry environment. No water gets into the mining entries.	Seasonal dry and wet environment. Some water into entries but is manageable.	Generally wet environment. Some water problems could be expected at fault zones.	Very wet environment. Water gets into entries from the pit and from the ground. It may flood the entry.	

(Continued)

Table 6.3 Continued.

Rank	Parameter	Condition				
		Excellent (100 ≥ rating > 80)	Good (80 ≥ rating > 60)	Fair (60 ≥ rating > 40)	Poor (40 ≥ rating > 0)	No-go (rating = 0)
C1	Coal strength	Lab coal UCS >30 MPa. Limited coal joints with wide spacing.	Lab coal UCS = 20–30 MPa. Coal joints with moderate spacing.	Lab coal UCS = 10–20 MPa. Coal joints with close spacing.	Lab coal UCS <10 MPa. Heavily cleated. Coal joints with very close spacing.	
C2	Stone bands	None, or fresh, weak to strong stone or dirt bands well fused at coal contacts.	Fresh, weak to moderately strong stone bands with rough surfaced partings anywhere across vertical hole diameter.	Fresh to weathered, weak dirt or stone bands with smooth to polished partings in top or bottom third of vertical hole diameter.	Weathered very weak, sheared, polished or slicked dirt or stone bands occurring in the top or bottom third of the vertical hole diameter.	Hard stone bands >0.2 m or moisture-softened dirt bands or clay seams sensitive to moisture occurring anywhere across the hole diameter.
C3	Cover depth (overburden + spoil)	Overburden depth 50–80 m in the reserve.	Overburden depth 70–100 m.	Overburden depth 80–120 m, possible stress concentration near highwall.	Overburden depth >120 m, possible significant stress concentration near highwall.	
C4	Seam thickness	1.5 m to 2.5 m	1.0 m to 1.5 m 2.5 m to 3.5 m	3.5 m to 5.0 m	>5.0 m	<1.0 m

Group B – Sub-critical parameters
  B1 Seam continuity
  B2 Pit condition
  B3 Seam quality
  B4 Groundwater
Group C – Important parameters
  C1 Coal strength
  C2 Stone bands in coal seam
  C3 Cover depth
  C4 Seam thickness

Guidelines for rating each of the 12 parameters are given in Table 6.3. Most of them are kept the same as used for HMI in Table 6.1. New ratings are given to the seam thickness and mineable reserve.

Similar to the HMI, the AMI is a weighted mean value of all 12 parameters. It is calculated using the following equations:

$$\overline{A} = (A1 \cdot A2 \cdot A3 \cdot A4)^{1/4}$$

$$\overline{B} = (B1 \cdot B2 \cdot B3 \cdot B4)^{1/4}$$

$$\overline{C} = (C1 \cdot C2 \cdot C3 \cdot C4)^{1/4}$$

$$AMI = (\overline{A}^{1.5} \cdot \overline{B}^{1.0} \cdot \overline{C}^{0.5})^{1/3}$$

where an exponential weight factor of 1.5, 1.0 and 0.5 respectively is given to groups A, B and C.

For each parameter, a confidence level is assigned. The confidence level of AMI is a weighted mean value of the confidence levels of all 12 parameters. The same weight factors used above in the AMI calculation apply to the calculation of the confidence level of AMI.

After both the AMI and the confidence level of AMI are estimated, the pit studied can be plotted in an AMI confidence level chart; see Figure 6.3. Depending on where the pit is positioned in the chart, a judgement can be made on the pit for auger mining.

6.5 SITE INVESTIGATION REQUIREMENTS AND METHODS

Following identification of a potentially suitable site for highwall mining, a complete investigation of a Highwall Mining Reserve (HMR) should achieve the following:

- Establish the geological viability of a proposed HMR;
- Design pillars that will perform adequately for the desired penetration depth and evaluate span stability;
- Refine the reserve geometry, identify gas conditions, and provide more detailed coal quality information; and
- Undertake a risk analysis to identify and manage specific risks arising from site-specific highwall and lowwall conditions.

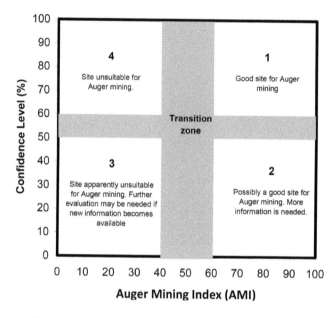

Figure 6.3 *Figure 6.3* AMI and confidence level chart for auger mining.

A detailed description of how to undertake an investigation for highwall mining is given in Duncan Fama et al. (1997) and that report has formed the basis for much of this section. A flow chart summarising the process is given in Figure 6.4.

6.5.1 Defining the highwall mining reserve

Initially, a Highwall Mining Reserve (HMR) needs to be established. This is done by considering the length of highwall that is available, the seam thickness, likely pillar widths and the probable penetration depths that will be achieved using the equipment that will be available for mining. A CHM system can potentially achieve penetration depths of up to 500 m, whereas auger systems can potentially achieve maximum penetration depths of about 225 m. Experience has shown, however, that these potential depths have often not been achieved, mainly because of unforeseen geotechnical conditions.

A first-pass definition of an HMR can usually be achieved by building a simple geological model based on a review of existing borehole logs, geophysical data, geological maps and sections and topographical information. Specifically, at this stage, the data should be reviewed to assess roof and floor conditions, the presence and potential impact of geological structures including faults and joints, whether any igneous structures are present and any gas impacts.

6.5.2 Seam-scale geology

Once a potential HMR has been identified, the site scale geology needs to be more closely defined – in particular, gross lithology, joints, faults, shears and igneous intrusions need to be identified. Similarly, seam thickness variability, seam rolls and

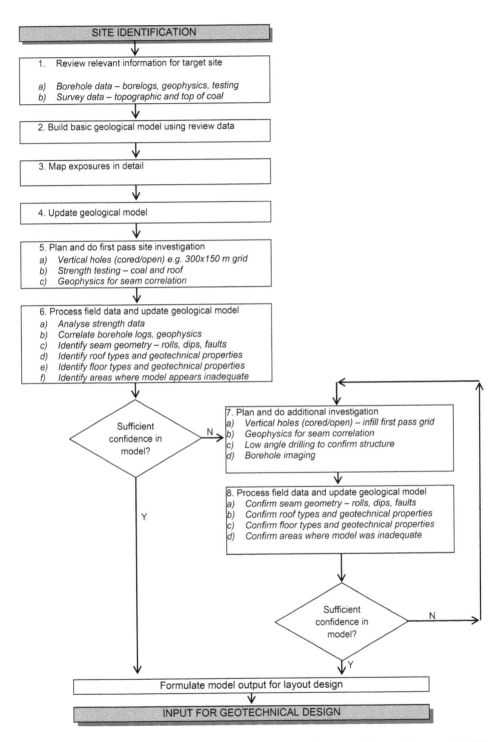

Figure 6.4 Flowchart showing the site investigation process and activities (Duncan Fama et al., 1997).

seam characteristics need to be identified as each of these factors can have a major impact on the amount of coal that is recoverable. The possibility of the presence, composition and concentration of any seam gas should also be investigated. If gas is present in too large a quantity to be adequately managed, it is likely to be a project stopper.

Following a review of existing information, the next level of required data is most easily gathered by a geologist mapping the proposed highwall. Mapping will provide a lot of data quickly. However, it must be remembered that these data represent only one slice of the total HMR and may not be representative of the overall geological conditions.

Following surface mapping, a geotechnical investigation is required, involving drilling that targets particular features and seeks to answer specific questions in relation to the proposed highwall mining.

6.5.2.1 Sub-surface investigations

Seam conditions

Thickness variability, roof washouts, seam dip and variability, seam rolls, folds, faults, igneous intrusions and other depth-limiting features need to be identified. For auger mining the critical factor is the limited ability of single auger flights to bend through sharp changes in seam dip. RMCHM (Rigid Modular Continuous Highwall Miners) are even more sensitive to changes in seam dip and are at risk of becoming entrapped.

Seam continuity can be investigated by vertical open-hole drilling and using geophysical logging to identify the roof and floor of the coal seam. These techniques are cheap compared with low angle drill holes and borehole imaging systems which may be required in some circumstances. From these data, plots of seam thickness and roof and floor dip may be generated to identify where seam inflections are occurring.

Seam strength may be inferred from existing strength test results determined on samples from the same seam combined with an evaluation of seam brightness profiles from within the proposed HMR. Where there are no existing laboratory strength results, strength tests should be carried out mainly on HQ (61 mm) core samples but supplemented by some tests on 100 mm samples to evaluate scale effects on coal mass strength.

Brightness profiles, cleat orientation, and the presence of partings or dirt bands can all be evaluated from core samples and mapping of the highwall face.

Roof and floor conditions

An investigation for highwall mining should establish:

- The rock types in the roof and floor, the thickness of the strata and their variability.
- Geological structures in the roof and floor. As a minimum, the pattern of joints in the immediate roof (i.e. to a distance of 5 m from the top of the coal) needs to be established, including orientation, spacing and persistence. The orientation, spacing and persistence of bedding partings and any infill materials should also be established for both the roof and floor.
- The strength of the roof and floor. As a minimum, Uniaxial Compressive Strength (UCS) should be inferred from a relationship established between sonic velocity and UCS. These results should be supplemented by the results from a large number

of point load strength index tests and a smaller number of good-quality uniaxial compressive strength test results. In this regard it should be noted that strength test results are highly dependent on rock moisture content and allowing core samples to dry out for even just a few minutes will lead to a substantial overestimate of rock substance strengths. Where it appears that the floor strength will be low and therefore have an impact on pillar performance, triaxial strength testing should be carried out so that a proper evaluation of bearing capacity may be made for design purposes.

Each of these parameters is important for the design of the highwall mining layout and taken together permit inferences about the values of other parameters such as Poisson's ratio and Young's modulus, which are necessary for numerical modelling.

Seam gas
Although outside the scope of this section, it is usual to check for seam gas during the course of a highwall mining field investigation. Excessive volumes of methane contained within the coal seam to be exploited can be a project stopper and so the seam gas content needs to be determined early in an investigation. This work is normally carried out by specialist contractors.

6.5.2.2   HMR-scale geology

The sub-surface investigation also needs to provide information on the scale of the whole HMR for formulation of a geological model that is adequate to help predict likely mining conditions.

Lithology
The rocks throughout the HMR should be described in terms of their type, stratigraphy, distribution, strength and weathering.

Bedding
Bedding should be identified as to style (e.g. is there cross bedding present in any sandstone strata), orientation, structure (e.g. is there any evidence of bedding plane shearing) and mechanical properties (e.g. are bedding planes open and therefore likely to form release surfaces or are they cemented and effectively of the same strength as the rock substance).

Jointing
The investigation should establish the main joint sets present in the HMR and provide information to describe for each set their orientation, spacing, persistence and termination styles. Each of these four parameters should be described at least semi-quantitatively so that at least an average value and a range are estimated. In addition, the relationship between joint intensity and proximity to faults should be investigated.

Faulting
The style and displacement of any faulting should be identified by the investigation. At the seam scale, faulting may be responsible for sterilisation or dilution of reserves. At the HMR scale, faulting may be a hazard due to increased risk of highwall instability,

the formation of release surfaces at a larger scale than local roof falls and an increase in the intensity of jointing in the vicinity of any faults. A number of studies have shown that old faults may be re-activated by mining and that even very local re-activation is sufficient to result in an increased incidence of roof fall.

Igneous intrusions

Dykes and sills are common occurrences in coal strata and their presence should be identified by a site investigation. The most common effects of igneous intrusions are: cindering of coal; weathering of the intrusive rocks, and sometimes their immediate host rocks, to materials having the strength of soil; and acting as aquifers or aquicludes, depending on their specific properties.

Extrusive igneous rocks

Extrusive igneous activity in a number of areas has resulted in basalt flows burying ancient soil profiles in former valleys with various impacts on highwall mining. Where the extrusive rocks are close to the coal, often the coal is oxidised to the extent of being valueless, and it and the surrounding rocks may be much reduced in strength.

Extrusive rocks are often highly fractured and consequently are water charged. When mining takes place, roof collapses induced by the pressures due to perched water tables may occur.

In situ stress

Generally, *in situ* stress, apart from overburden stress, does not pose a problem to highwall mining. The depths at which highwall mining is carried out are usually sufficiently shallow that most excess *in situ* stress has been dissipated.

Nevertheless, there are circumstances in which *in situ* stress may be sufficiently high to have an impact on mining. These circumstances include mining in steep terrain where *in situ* stresses may have been locked in, and mining in areas that have been affected by mountain-building processes or volcanic intrusions in geologically recent times. In the established coal-mining regions of Australia there is an adequate database of information to indicate whether or not *in situ* stress is likely to be a problem. If further investigation of *in situ* stress is required, it is usually carried out by specialist contractors.

Groundwater

Excessive groundwater can cause an increased incidence of roof failure as a result of the pressure acting on free faces exposed by the mining process. If the problem occurs, it is usually only for a short period until the groundwater pressures dissipate as mining progresses. It is routine for drillers to note groundwater levels and any unusual flows on their logs and this should provide a sufficient guide to evaluate the likely impact of groundwater.

Overburden

The overburden conditions need to be investigated to define adequately the cover load. The following issues need to be addressed:

- Density of the overburden rock types;
- Spoil height, location and geometry; and
- Influence of benched highwalls.

In the case of critical layouts such as fanned layouts, it may be necessary to carry out numerical analysis of the impact of the overburden on the proposed layout.

6.5.2.3   Highwall mining hazards

Just before mining commences it is normal practice to undertake a risk analysis in order to minimise the impact of any potential hazards.

Highwall hazards
Highwall hazards are generally evaluated by mapping the highwall, systematically identifying hazards and providing recommendations on how to minimise them. The scheme derived for BHP Coal is shown as an attachment together with an example of how the results of the risk analysis can be reported.
The main issues to be investigated are:

* Loose rocks on the face or crest;
* Soil or weathered rock slumping from the crest;
* Potential wedge, block or toppling failures;
* Any effects of blast damage; and
* Undercut horizons.

Lowwall hazards
The main hazard to be investigated is deep and shallow lowwall spoil stability, including falling or rolling boulders.

Other hazards
As part of the geotechnical investigation for the risk analysis, consideration should also be given to the possible effect of high wind – loose rocks, perhaps surprisingly, are often dislodged and rainstorms may induce mass movement as well as flood the pit in which the highwall mining activities are being carried out.

6.5.3   Conclusion

This section has provided an outline of the site investigation requirements and a description of the methods that may be used to achieve those objectives. Site investigation for a highwall mining reserve should follow a logical sequence in which existing data are reviewed, specific questions are framed and targets identified for the site investigation, and then the results of the investigation should be carefully reviewed. If the questions have not been answered or if critical targets have not been reached then a second stage of investigation may be necessary, using more sophisticated and more expensive tools, to resolve the outstanding issues.
Following completion of the site investigation there will be a detailed geological model for the HMR, parameters for the geotechnical design of the HMR will have been established, and a risk analysis should have been carried out.

6.6 SPAN STABILITY ASSESSMENT

This section outlines the assessment procedure for span stability of highwall mining entries. The recommended procedure for unsupported span stability assessment involves three main steps:

1 Identify the span failure mechanisms;
2 Select proper tools for span stability assessment;
3 Assess the stability of entry spans using all of the available geological and geotechnical data.

Each of the three steps is discussed in detail below.

6.6.1 Unsupported failure mechanism

Unsupported span failure mechanisms are controlled by a number of factors, including rock strength, *in situ* stresses, joints, lithology, sedimentology, and excavation geometry. In general, failure mechanisms can be divided into five main types, as follows.

6.6.1.1 Span delamination and snap-through

Span failure starts with delamination, followed by tensile cracking of the immediate roof layer, which finally breaks due to snap-through. This type of failure often occurs in roofs with weak inter-lamination strength and low to moderate *in situ* stresses (Figure 6.5).

6.6.1.2 Coal beam shear failure

When the immediate roof is a layer of cleated coal, it may fall from shear failure along existing vertical cleats. Lack of horizontal confinement and loading from overburden rock mass contribute to this type of failure (Figure 6.6).

6.6.1.3 Block fall

This type of span failure often occurs in strong roof rocks where joints are well developed. In this case, joint distribution and joint strength dictate span stability (Figure 6.7).

6.6.1.4 Intact rock shear failure

This mechanism often occurs where the roof rock is weak but relatively intact and the *in situ* stresses are high in comparison with the rock strength. Spans may fail due to this mechanism in underground roadways and, sometimes, highwall mining entries in a high horizontal and vertical stress regime (Figure 6.8).

6.6.1.5 Span buckling

Buckling failure occurs when the horizontal stress is relatively high, the roof strata are thinly laminated and the inter-lamination strength is low (Figure 6.9).

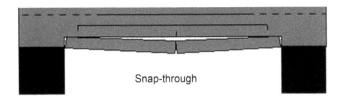

Figure 6.5 Span instability caused by roof delamination and snap-through.

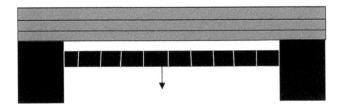

Figure 6.6 Coal roof instability caused by shear failure along coal cleats.

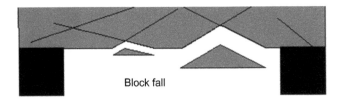

Figure 6.7 Span instability caused by block falls.

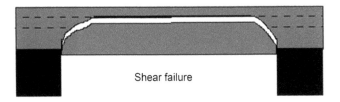

Figure 6.8 Span instability caused by shear failure.

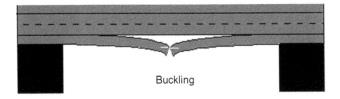

Figure 6.9 Span instability caused by buckling.

For typical Australian highwall mining conditions (laminated roof with low to moderate *in situ* stresses), the most likely failure mechanisms are the snap-through and coal beam shear mechanisms. Joint-controlled block falls are also common in areas where intensive jointing has occurred in the immediate roof.

Two analytical models have been developed (Shen & Duncan Fama, 1996) for the assessment of span instability due to the snap-through and coal beam shear failure mechanisms. Block fall failure mechanisms should be analysed using numerical tools such as *UDEC* (Itasca, 1996) and *3DEC* (Itasca, 1994). The shear failure mechanism and the bulking failure mechanism are not very common in highwall mining conditions. They may be assessed by using existing numerical codes for continuous or discontinuous media.

6.6.2   Span stability assessment tools

6.6.2.1   The Laminated Span Failure Model (LSFM)

The LSFM was developed by CSIRO specifically for highwall mining span stability assessment in the thinly laminated roof strata that often occur at highwall mining sites in the Bowen Basin of Queensland. The LSFM models the roof as a series of thin rock beams which are detached from each other and which can progressively fall out of the roof. The LSFM combines an equation to calculate the maximum stable span with a simple technique to estimate the likelihood of span failure and the height of fall for a given highwall mining span.

In two dimensions, the LSFM considers the immediate roof layer as a thin and long beam which is detached from the overlying rock. Cracks can develop near the abutments and at the mid-span as the beam deflects under its self-weight. The final failure of the immediate roof beam is caused by snap-through at the mid-span; see Figure 6.5. According to the LSFM, the maximum stable span is controlled by the roof layer thickness, Young's modulus and compressive strength, and load applied to the layer (Shen & Duncan Fama, 1996). The maximum stable span is easily calculated using a simple closed-form solution as follows:

$$S = 1.873 \cdot t \cdot \left(\frac{E}{p}\right)^{1/4} \cdot \delta\left(\frac{\sqrt{Ep}}{\sigma_c}\right) \tag{6.1}$$

where,
S – span (m)
T – roof layer thickness (m)
E – Young's modulus (MPa)
p – distributed vertical load on the roof layer (MPa), in most cases being the self-weight of the layer. However, it can be greater due to additional load from the overlying layers, particularly when the overlying layers are not horizontal.
δ – a dimensionless factor to consider rock crushing (≤ 1.0). It is a function of the compressive strength, Young's modulus of the roof rock, and the load.
σ_c – unconfined compressive strength (MPa).

To take account of the likely variability in roof conditions across a highwall mining reserve, the LSFM incorporates a simple technique to assess the probability of span

Table 6.4 Young's modulus values and their assigned probabilities.

Low Value		Most Likely Value		High Value	
E_1	$p(E_1)$	E_2	$p(E_2)$	E_3	$p(E_3)$
5 GPa	0.2	8 GPa	0.6	11 GPa	0.2

instability. This is required because a highwall mining reserve (HMR) typically extends over an area of 0.3 to 0.5 km². Roof conditions will vary over an area of this magnitude. One way to account for this variation is to define a range of possible values for each parameter and assign a probability to the high, low and most likely value. For instance, if we know that the Young's modulus of the roof rock is likely to be in the range 3–12 GPa and considered most likely to be 8 GPa, we may define three values for this parameter (Table 6.4).

The probability, $p(E_i), i = 1, 2, 3$, assigned to each value, represents a subjective judgement as to its likelihood of occurring based on the best available information. This can be adjusted and the analysis re-run if and when new information becomes available.

All the input parameters used in Equation (6.1) can be defined in this way. Using basic statistical theory, the assessment results can be presented in a probabilistic way. Typically, results might be shown as a graph of span width versus the cumulative probability of span failure (Shen & Duncan Fama, 1996).

The height of possible roof falls for a given span is also estimated in the LSFM. It is assumed that the failure tends to create an elliptical opening whose shape is determined by the ratio of the horizontal and vertical stresses. Details of the calculation are described in Shen and Duncan Fama (1996).

6.6.2.2 The Coal Roof Failure Model (CRFM)

In many cases, a coal layer is left in the immediate roof either because of a limitation in the mining height or with the aim of improving span stability. When roof coal is left, it is the stability of the coal that is the issue and not necessarily the stability of the non-coal strata. The mechanism of coal falls is notably different from that of laminated roof rocks, due to the structural differences between the two materials. Rock falls often occur as thin slabs, but roof-coal failures often incorporate the whole thickness of the unmined coal. The dominant mechanism in coal roof failures is shear failure along sub-vertical cleats (Figure 6.6), which is often associated with low horizontal stresses in the coal seam.

A typical failure mechanism of roof coal is shown in Figure 6.10. After the excavation of an entry, the roof-coal beam deflects in response to the vertical stress from the main roof, while the main roof deforms in response to stress redistribution in the vicinity of the entry. The coal beam often deflects in a different manner from the main roof, and, when it does, the coal beam separates maximally from the main roof in the central part of the span. However, contact between the coal beam and the main roof still exists near the ends of the span. There, the coal beam is loaded mainly vertically, through the contact with the main roof. The vertical load on the beam leads to high

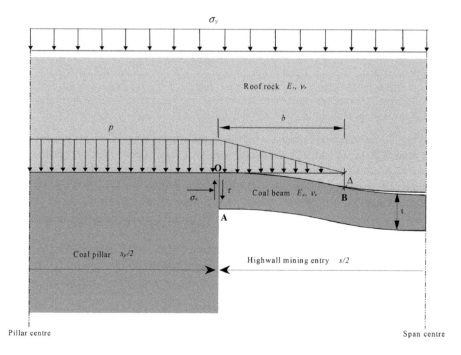

Figure 6.10 A coal roof failure model.

shear stress in the beam cross section at the ends of the span. If the resultant shear stress is high enough and the horizontal confinement is low enough, shear failure will occur and the whole coal beam will fall into the entry.

A formula for this type of failure has been derived. The details are given in Shen and Duncan Fama (1999a). In summary, the Factor of Safety (FoS) of the coal beam is given by the equation below:

$$FoS = \frac{\left(\sigma_x + \dfrac{b(s + s_p)\sigma_y \tan(\phi_i)}{2t(b + s_p)} \dfrac{(s - 2b/3)}{(s + s_p)}\right) \tan\phi + c}{\dfrac{1}{2}\dfrac{b}{t}\dfrac{s + s_p}{b + s_p}\sigma_y} \tag{6.2}$$

where,

ϕ, c – friction angle and cohesion of sub-vertical coal cleats;

σ_x – *in situ* horizontal stress in coal seam;

σ_y – overburden stress;

ϕ_I – friction angle of coal/roof interface;

s, s_p – entry width and pillar width;

t – thickness of roof coal;

b – contact length between the coal beam and roof rock.

The contact length b is the solution of the following equation:

$$\frac{2(1-\nu_r^2)}{\pi}\frac{b}{t}\left[\frac{3}{2}-\ln b+\ln s_p\right]-\frac{E_r}{E_c}\left[\frac{b^4}{10t^4}\left(4-5\frac{b}{s}\right)+\frac{(1+\nu_c)}{3}\frac{b^2}{t^2}-1\right]=0 \quad (6.3)$$

where E_r/E_c is the stiffness ratio of the roof rock (E_r) and coal (E_c); ν_r and ν_c are Poisson's ratio of the roof rock and coal.

An example of the assessment of coal roof stability is given in Section 6.6.3.2.

6.6.2.3 Numerical tools

If the span failure is caused by a combination of two or more different mechanisms or the geology is too complicated to use the analytical model, then numerical tools may be used. The numerical methods used to analyse roof stability can be classified into two categories: the continuum approach and the discontinuum approach.

In the continuum approach, roof rocks are modelled as elasto-plastic materials. Intact rock yielding is considered to be the indicator of roof failure. There are many finite element and finite difference codes which use the continuum approach. The most commonly used codes in stability assessment of underground openings are *FLAC* and its three-dimensional version *FLAC3D* (Itasca, 1995, 1996).

When using *FLAC* to predict the stability of entries, the rock mass is treated as a continuous medium. Bedding planes can be simulated using the ubiquitous joint model, which considers possible failure along the bedding planes without involving intact rock. The advantage of using *FLAC* for span stability prediction is the simplicity of the numerical model and the calculation speed. However, *FLAC* modelling does not predict roof falls explicitly but only provides an indication of the extent of roof plasticity. In an actual mining span, plastic yield in the roof does not necessarily lead to roof falls because the yielded roof rocks can still be held in place by horizontal confinement and friction.

Recently CSIRO has developed a Cosserat continuum code for span stability assessment. The Cosserat model (Cosserat & Cosserat, 1909) provides a large-scale (average) description of a layered medium. In this model, inter-layer interfaces (joints) are considered to be smeared across the mass, i.e. the effects of joints are implicit in the choice of stress-strain model formulation. An important feature of the Cosserat model is that it incorporates bending rigidity of individual layers in its formulation and this makes it different from other conventional implicit models such as the ubiquitous joint model (Itasca, 1995), which is based simply on strength anisotropy theory. Thus the code considers the presence of bedding planes in the roof by including the effect of such bedding planes in its mathematical formulation. The code may also model rock creep and therefore is able to simulate time-dependent span stability. Application of the Cosserat code in underground roadway span assessment has shown promising results (Kelly et al., 2001).

In the discontinuum approach, the roof strata are considered to be an assembly of blocks or layers. Joints and bedding planes are modelled explicitly. Roof instability is caused by the fall of roof blocks or the collapse of roof layers. The numerical codes used to analyse discontinua include distinct element codes such as *UDEC* and its 3D version *3DEC* (Itasca, 1996, 1994) and the particle flow codes (Sakaguchi et al., 1999; Itasca 1999).

UDEC is the most commonly used discontinuum code for span stability assessment. It requires detailed knowledge of a range of geological and geotechnical parameters. These parameters include:

- joint distribution, joint strength
- bedding spacing, bedding strength
- intact rock strength and deformability
- *in situ* stresses

Experience has shown that *UDEC* models are sensitive to the values selected for the input parameters. Consequently, considerable numerical modelling experience is required to properly set up and run a model and to explain properly the results that are produced.

6.6.2.4 Empirical tools

A number of empirical tools have been developed to assess span stability but they do not distinguish between failure mechanisms. Most of the empirical methods are based on rock classification systems. There are three commonly used methods: Rock Mass Rating (RMR) system (Bieniawski, 1993), Q-system (Barton et al., 1974) and Coal Ming Roof Rating (CMRR) system (Molinda & Mark, 1994). Both the RMR system and the Q-system were developed for hard rock conditions and are not widely used in the coal-mining industry.

The CMRR system was developed by the U.S. Bureau of Mines as an engineering tool to quantify descriptive geological information for use in coal mine design and roof support selection. The CMRR weights the geotechnical factors that contribute to roof competence such as bedding spacing, jointing, and groundwater, and combines them into a single rating on a scale of 0–100.

Recent studies conducted in underground place change development headings have shown that the CMRR can be used to evaluate unsupported span stability (Mark, 1999). The results are summarised in Figure 6.11.

The equation for the "extended cut" line shown in Figure 6.11 (the upper line) is as follows:

$$CMRR_{crit} = 19.2 + 5.38 W_e \tag{6.4}$$

where W_e is the entry width in metres. Equation (6.4) separates the "sometimes stable" from the "always stable" groups. For a highwall mining entry width of 3.5 m (as fixed by the CHM cutting head), Equation (6.4) implies $CMRR_{crit} = 38$.

Equation (6.4) is based on data from roadway spans in underground coal mines. In a highwall mining entry, the overburden stress is much lower and the unsupported entry is much longer than that of the "extended cut" (which is often no more than 6–12 m) in an underground roadway. There is very limited data where the CMRR system has been applied in highwall mining span stability assessment. The only reported application to highwall mining was in a mine in Central Queensland. The CMRR system was used to assess the span stability of the 3.5 m roadway and the prediction results did not

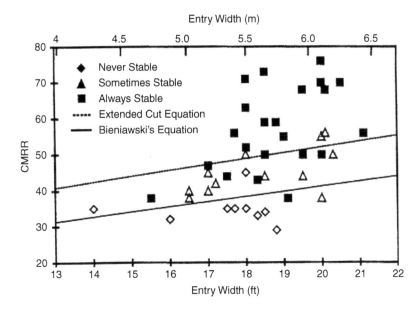

Figure 6.11 Relationship between CMRR and entry width for unsupported spans (Mark, 1999).

agree with the actual span performance, possibly due to the lack of accurate geological information. Accordingly, using the CMRR system in highwall mining should be done cautiously.

6.6.3   Case studies

For a given site, the most likely span failure mechanism(s) should be initially identified. Prediction tools suitable for the particular failure mechanism(s) are then selected. Depending on the type of tool selected, a range of geological and geotechnical data are needed for the analysis. These data should be obtained from the site investigation and laboratory testing. Once all the necessary input parameters are known, the assessment can be carried out. Three case studies are presented which demonstrate how the prediction tools can be applied to the highwall mining span stability assessment. The procedure adopted in each case was to identify the most likely span failure mechanism(s), select the appropriate tools to analyse the mechanisms and then analyse the mechanisms using the available geological and geotechnical data.

6.6.3.1   Case study 1

Highwall mining was to be carried out in a pit in Central Queensland. The whole seam thickness was to be cut during mining so that the immediate roof rock would be exposed in the entry spans.

The immediate roof (<5 m) was mainly composed of laminated siltstone. The thickness of the siltstone laminae ranged from 20 mm to 300 mm. The roof rock was weak (UCS = 10–40 MPa) and the roof bedding was slightly angled (<15°) across the

Table 6.5 Mechanical parameters of roof strata in an example highwall mining pit.

	Lowest	Low	Intermediate	High	Highest
E (MPa)	5000		10000		20000
Probability	0.25		0.5		0.25
t (m)	0.025	0.05	0.1	0.17	0.25
Probability	0.1	0.2	0.4	0.2	0.1
p	1.0 × self-weight		1.5 × self-weight		2.0 × self-weight
Probability	0.8		0.15		0.05
σ_c (MPa)	15		25		35
Probability	0.3		0.5		0.2

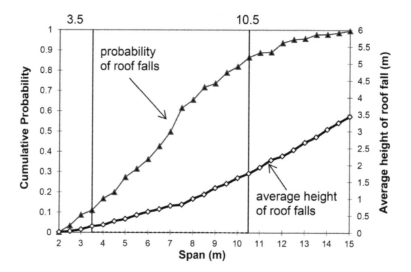

Figure 6.12 Predicted probability of roof falls and the average height of the falls using the Laminated Span Failure Model (LSFM), case study 1.

entry spans. The overburden depth ranged from 60–80 metres. Joints were widely spaced in the roof rock.

It was assessed that the most likely span failure mechanism would be delamination and snap-through. Therefore, the analytical snap-through model (LSFM) was used to assess the span stability.

Based on the information from a site investigation and from laboratory test data, the input parameters for LSFM and their probabilities over the area of highwall mining reserve were estimated and are listed in Table 6.5.

Using Equation (6.1) and the probabilistic approach described in Shen and Duncan Fama (1996), the probability of roof falls and the average height of the falls were obtained for a range of span widths (2–15 m), and are shown in Figure 6.12.

For an ordinary highwall mining entry with a span width of 3.5 m, the model predicted an 11% chance of roof falls. The average height of the falls is predicted to be 0.2 m. The chance of roof falls increases with increasing span width, as does the height of the roof falls. For instance, with a 10.5 m span, the probability of a roof fall

Table 6.6 Parameters selected as input to the model.

Span width (s)	3.5 m
Pillar width (s_p)	3.7 m
Roof coal thickness (t)	1.0 m
Young's modulus of roof rock (E_r)	3 GPa
Poisson's ratio of roof rock (v_r)	0.25
Young's modulus of coal (E_c)	2 GPa
Poisson's ratio of coal (v_c)	0.23
Friction angle of coal cleats (ϕ)	41°
Cohesion of coal cleats (c)	1.1 MPa
Friction angle of roof/coal interface (ϕ_i)	16°

is 86% and the average height of fall is 1.8 m. This extent of roof fall is considered to be severe, and would likely damage, and perhaps bury, the mining equipment.

The actual mining records from this pit show that some minor roof falls did occur during mining, mainly as thin roof slabs. On one occasion, a 10–11 m span was created by the convergence of three entries owing to poor mining guidance. A roof fall extending to a height of about 2 m occurred. The observed span failure agrees well with the prediction.

6.6.3.2   Case study 2

A highwall mining pit in Central Queensland had a 4.8 m thick coal seam. Due to the limited cutting height capacity of the continuous miner to be used, a layer of coal, approximately 1.0 m thick, was to be left in the roof. The stability of the roof coal was a concern, as 1.0 m thick coal falls can disrupt mining operations.

The Coal Roof Failure Model (CRFM) was used to assess the stability of the coal layer. The following mechanical and geometrical parameters were selected for input to the model (Table 6.6). These values were based on the results of a site investigation of the pit.

The estimated overburden stress ranged up to 2.7 MPa over a penetration depth of 0–300 m for the analysis, and the maximum overburden stress of 2.7 MPa was used. The ratio of horizontal stress to overburden stress in the coal seam was not known, but it was considered that it could be as low as $\sigma_H/\sigma_v = 0.5$ based on limited stress measurement results in adjacent areas. Three cases of low to high stress ratio ($\sigma_H/\sigma_v = 0.5, 0.75, 1.0$) were investigated.

Using the CRFM model and Equation (6.2), the calculated Factors of Safety (FoS) for the three stress ratios were:

σ_H/σ_v	FoS
0.50	0.83
0.75	1.02
1.0	1.21

The results indicated that, if the stress ratio in this pit was 0.5, then the FoS of the coal roof would be well below 1.0 and unstable coal roof would almost certainly result. In contrast, if the stress ratio was 1.0, the FoS of the coal roof would be about 1.2 and this would likely result in a stable coal roof. No measurements of *in situ* stress had been carried out in this pit so no more definitive predictions could be made.

Table 6.7 Mechanical properties used for UDEC modelling of span stability.

	Young's modulus E (MPa)	Poisson's ratio ν	Unconfined Compressive Strength UCS (MPa)	Cohesion c (MPa)	Friction angle ϕ (°)	Tensile strength σ_t (MPa)
Coal	2000	0.23	4.8	1.09	41	0.1
Roof mudstone	3000	0.25	12.5	2.8	41	0.1
Floor mudstone	2000	0.25	4.05	0.93	41	0.2
Roof siltstone	5000	0.25	(elastic)			
Sandstone	10000	0.25	(elastic)			
Coal cleat				0.5	30	0
Coal joint				0.55	41	0
Mudstone bedding				0.1	26	0
Coal/roof/floor interface				0.1	26	0
Rock joint				0	35	0
Rock/rock interface				0.5	35	0

During the actual mining of this pit, extensive roof falls were experienced, from which it was inferred that low *in situ* horizontal stresses existed in the pit.

6.6.3.3   Case study 3

A pit in Central Queensland was to be highwall mined. The coal seam thickness was 2.1 m. Very closely spaced angled coal joints were found in the coal seam. The coal seam was overlain by a layer of 0.8 m thick, weak, laminated mudstone. It was expected that the mudstone layer would fall and cause problems during mining. It was thought that leaving coal in the roof might improve the roof conditions. An assessment was required as to how much coal should be left in the roof to ensure the span stability.

This case involved the stability of both jointed roof coal and overlying laminated mudstone. Consequently, the numerical code *UDEC* was selected to analyse the case owing to its flexibility in handling complex geology. *UDEC* models require comprehensive geological and geotechnical data. The mechanical properties relevant to the pit studied are listed in Table 6.7. The values of the properties were estimated based on the results of site investigation and previous modelling experience.

One highwall mining entry and two half pillars, one at each side of the entry, were modelled. The entry was excavated in one step. This model simulated the ultimate span stability when highwall entries at both sides of an entry were excavated. The bedding in the roof mudstone, sub-vertical joints in the roof and inclined joints in the coal seam were all modelled explicitly. The roof bedding thickness was modelled to vary randomly with an average thickness of 0.1 m. Two cases were studied where roof coal thickness was 0.3 m and 0.5 m respectively.

An overburden depth of 70 m was modelled. The horizontal/vertical stress ratio was assumed to be 0.5 in the coal seam and 1.0 in the roof rock.

The modelling results for the two cases with different coal thickness are plotted in Figures 6.13 and 6.14. When the roof coal was 0.3 m thick in the model, the results

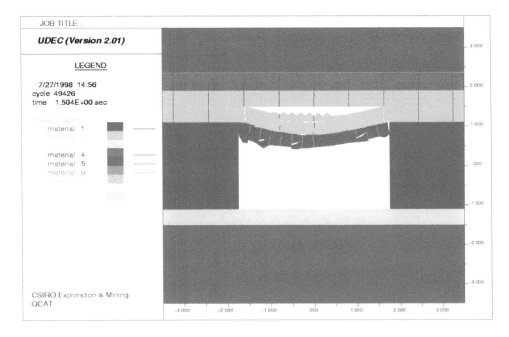

Figure 6.13 Modelling results – unstable span with 0.3 m coal roof.

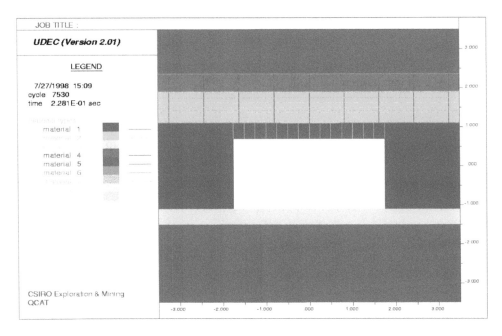

Figure 6.14 Modelling results – stable span with 0.5 m coal roof.

predicted that the coal layer and 0.4 m of roof mudstone would fall. When the roof coal was 0.5 m thick in the model, a stable span was predicted.

During the actual mining operations, a coal layer was left in the roof of the first two entries in the pit. The thickness of the roof coal, however, was difficult to control, and it ranged from 0.2 m to 0.5 m. Seven coal falls were reported during mining of these two entries. After that, no roof coal was left in the entries and the weak roof mudstone was cut during mining. The rock roof, dominated by stronger siltstone, was mostly stable.

6.7   LAYOUT DESIGN

Layout design tools fall into two main categories, namely numerical modelling methods and empirical techniques. Both rely on sound geological inputs. Section 6.7.1 describes the CSIRO numerical modelling technique, which is used by CSIRO for most new pit layout designs. Classical empirical techniques are summarised in Section 6.7.2 and a new empirical curve for estimation of the strength of highwall mining pillars is proposed. Based on the latter, there follows a calculation of preliminary estimates of pillar widths and percentage extraction for given inputs.

6.7.1   The CSIRO numerical modelling method

6.7.1.1   Summary of numerical modelling tools

The following three numerical codes have been used for the CSIRO pillar design technique:

1   *FESOFT*, a CSIRO Finite Element Code (Duncan Fama et al., 1995)
2   *FLAC*, a Finite Difference Code (ITASCA, 1995)
3   *UDEC*, a Discrete Element Code (ITASCA, 1996).

A detailed description of the computer models and their usage by CSIRO, as well as the Hoek-Brown pre- and post-failure yield criteria as used by CSIRO for modelling coal pillars, is given in the reference cited above with additional detail given in Duncan Fama et al. (1999b). The most commonly used code outside CSIRO is *FLAC*, and each user has his own method of setting up a model and choosing appropriate inputs.

6.7.1.2   Mechanism of failure of a single pillar

Pillars are assumed to fail from the application of excessive compressive stress. The strength of the coal is governed by a Hoek-Brown yield criterion with post-peak softening characteristics. As increasing stress is applied to the pillar, yield zones grow from the edges – physically this causes rib spalling – but the pillar remains stable as long as an elastic core is maintained at the centre of the pillar. Once the yield zones coalesce, the pillar stress-carrying capability reduces and with increasing applied stress the pillar strength decreases rapidly to its residual value.

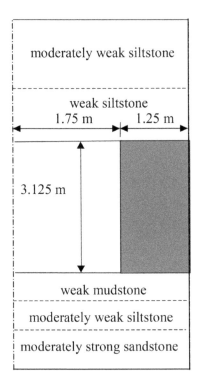

Figure 6.15 Geometry of the pillars and entries as modelled (not to scale vertically).

6.7.1.3 Example of stress-strain behaviour of a single pillar

The following section is summarised from Adhikary and Duncan Fama (2001). For a full description of model inputs and procedures the reader is referred to this report.

A two-dimensional plane strain model with just half of a 2.5 m wide pillar and half of a 3.5 m wide and 3.125 m high entry was constructed, as shown in Figure 6.15. The model had roller boundaries on three sides (i.e. left, right and the base), which implied symmetry conditions there. Effectively an infinite row of pillars and entries in the horizontal direction was being modelled. For simplicity, in the initial study, it was assumed that the interface between the coal and roof or floor rock was bonded.

The input parameters for the model are those commonly used for the D Upper seam at Moura Mine, where the coal strength (mass UCS = 4.8 MPa, cubic strength = 6.36 MPa) is similar to average-strength South African coal. The model was subjected to a continuously increasing displacement on the top boundary. The stress-strain plot obtained from the analysis is presented in Figure 6.16. Note that the linear section of the curve, where the axial stress ranges from 0 to about 5 MPa, is where there is virtually no yield in the pillar, even in the ribs. As the strength increases from about 5 MPa up to its peak value of 6.4 MPa, the yield zones grow from the ribs towards the core, but do not coalesce. In general, at peak strength, nearly one-third of the pillar is still not yielded. As the yield zones increase and eventually coalesce,

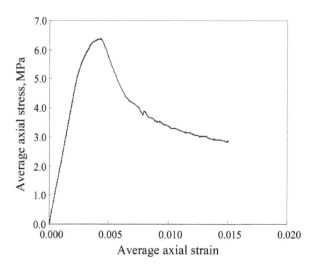

Figure 6.16 A stress-strain plot of 2.5 m wide coal pillar.

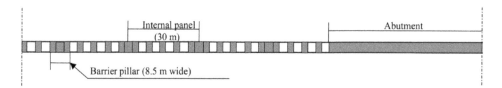

Figure 6.17 A schematic of a panel layout with four out of five entries mined and abutments on either side (seam thickness 3.1 m, pillar width 2.5 m, entry width 3.5 m).

so that the pillar is yielded right through, the axial stress decreases and moves down the post-failure curve.

6.7.1.4   Layout design

Figure 6.17 shows a typical panel 261 m wide, consisting of nine sub-panels (internal panels) each of width 30 m, where only four out of each set of five entries is to be mined, and with 100 m wide abutments. For any given layout design, the layout strength is the maximum stress for which the layout remains stable. This is determined by gradually increasing the stress levels until all pillars are yielded through.

A factor of safety for the layout, defined as the ratio of layout strength to the actual estimated cover stress at any given penetration depth, is calculated. The reader is referred to Adhikary and Duncan Fama (2001), and also to Duncan Fama et al. (1999a), for analyses of panel layouts with and without inter-panel barriers. No general guidelines exist as to whether layouts should be designed with or without barrier pillars. Industry practice includes both. Barrier pillars are intended to isolate panels, so that a local pillar failure does not propagate over a large area. However, this requires a good knowledge of just how wide these barrier pillars must be in order to achieve that objective. No research has been done into this question, and conservative design is costly in terms of extraction percentage. In general, CSIRO designs have been for parallel layouts without barriers. This is because it has been found that these layouts return a greater extraction percentage for a desired factor of safety than layouts with barriers (Follington et al., 1994). Section 6.7.2 gives an indication of required pillar widths for parallel layouts without barriers, based on empirical techniques described below.

6.7.1.5 Example

As a detailed example, a direct link is provided here to the pillar design that was carried out for Pit 20D Upper, Moura Mine. The full report (Duncan Fama et al., 1999a), which includes site investigation as well as layout design, is available by kind permission of BHP Coal Pty Ltd and Peabody Mining Pty Ltd.

6.7.2 Empirical techniques

6.7.2.1 Existing empirical methods

Empirical pillar design techniques, based on tributary area theory, are widely used in the design of underground bord and pillar mines (Bieniawski, 1987; Salamon & Munro, 1967; Salamon et al., 1998). None of those formulae, which were based on data from bord and pillar workings, are directly applicable to highwall mining pillars. In a previous publication (Duncan Fama et al., 1999b), empirical formulae were tentatively derived, based on models not data. A different approach to an empirical design curve is proposed here. It also seemed appropriate to bring any attempts at highwall mining empirical design into line with the underground approaches.

As is the case for underground pillars, the extent of the loss of confining stress in the pillar, after it is formed, plays a major role in pillar strength. Very narrow pillars lose almost all their confining stress. As well as the influence of the shape of the pillar, in particular the cross-sectional width to height ratio, the level of confining stress depends on the strength of the roof and floor and also on the presence or absence of sheared interfaces between the coal seam and the roof and floor.

In contrast to large width/height underground pillars, coal strength plays a major role in the pillar strength of the long, narrow pillars typical of highwall mining. This is a major difference which must be emphasised. Mass coal strength has a minimal effect on pillar strength for a squat underground pillar, but has a major effect on the strength of typical highwall mining pillars. It is not difficult to see why, from a consideration of the simple Mohr-Coulomb yield criterion, which defines the maximum principal stress, (σ_1), that any element in the pillar can sustain before yield:

$$\sigma_1 = k * \sigma_3 + \sigma_c \tag{6.5}$$

where σ_3 is the minimum principal stress in the element in the pillar, σ_c is the mass UCS (Unconfined Compressive Strength) of the coal and k is related to the angle of internal friction, ϕ, by $k = (1 + \sin\phi)/(1 - \sin\phi)$. If ϕ is 30°, $k = 3$.

If the coal UCS = 5 MPa, then at 500 m underground the vertical and horizontal stresses are about 12 MPa *in situ*. After a pillar has been formed by driving entries

Table 6.8 The effect of coal strength on yield stress at a point in a coal pillar.

Depth of cover	Roof/Floor	σ_3	σ_c	σ_1	% decrease in σ_1 when σ_c decreases from 5 to 3 MPa
500 m	Strong	12 MPa	5 MPa	41 MPa	−5%
500 m	Strong	12 MPa	3 MPa	39 MPa	
500 m	Weak	2 MPa	5 MPa	11 MPa	−18%
500 m	Weak	2 MPa	3 MPa	9 MPa	
80 m	Strong	1 MPa	5 MPa	8 MPa	−25%
80 m	Strong	1 MPa	3 MPa	6 MPa	
80 m	Weak	0.5 MPa	5 MPa	6.5 MPa	−31%
80 m	Weak	0.5 MPa	3 MPa	4.5 MPa	

on either side, a wide-enough pillar, with a strong roof and floor, may not lose any of this confinement in its core and $\sigma_1 = 41$ MPa. If the coal strength is 3 MPa, then $\sigma_1 = 39$ MPa. For practical purposes this difference is negligible – the UCS value ($\sigma_c = 5$ or 3 MPa) is insignificant compared with $k * \sigma_3 = 36$ MPa.

By contrast, at 80 m cover in a highwall mining panel, the *in situ* vertical and horizontal stresses are about 2 MPa. In general, this horizontal stress will not be retained in the pillar core, and may drop to about 1 MPa even with a strong roof and floor. Then $\sigma_1 = 8$ MPa when $\sigma_c = 5$ MPa, and $\sigma_1 = 6$ MPa when $\sigma_c = 3$ MPa. This difference is far from negligible. For shallower depths, or for weaker roof and floor, the differences are even greater. Table 6.8 shows these effects together with the possible effect of strong or weak roof or floor.

6.7.2.2 Proposed average empirical pillar design curve for highwall mining pillars

It is extremely important to note that the empirical approach proposed here will only be appropriate for panels where all entries are parallel and are mined over a panel of sufficient length that it can be assumed that full cover stress needs to be carried by the pillars in the centre of the panel. Where a series of smaller panels are designed, separated by barriers, analysis of stress carried by single pillars is not representative of the layout strength. This is because proportionally more stress will be carried on the barrier pillars.

Panels consisting of fanned entries will normally result in a series of smaller pillars separated by barriers, as described above. However, at full penetration depth, they may be designed so that the barriers reduce to the same width as the inter-panel pillars. In other cases, such as the Moura 17D upper central fanned layout, which had 40 entries, it is valid to assume full cover stress at the centre of the panel.

The empirical formula given in Equation (6.6), which is the "best-linear-fit" to the data, is proposed here for the peak strength of a highwall mining pillar of cross-sectional width W and height h, over a range of $0.5 \leq W/h \leq 2.0$ only. It is based on data from Duncan Fama et al. (1999b), with additional data from Adhikary and Duncan Fama (2001). This is for a coal seam of average strength, with roof and floor conditions typical of the D upper seam at Moura (moderately strong roof and roof/coal

Table 6.9 Peak strength of pillars according to Equations (6.6)–(6.9).

W/h	$S_{CSIROPART}$ (6.6)	$S_{BIENIAWSKI}$ (6.7)	S_{WAGNER} (6.8)	S_{CSIRO} (6.9)
0.5	4.82	4.42	6.00	4.92
0.71	5.29	5.38	6.92	5.38
1.00	6.36	6.00	8.16	6.27
1.29	7.43	6.63	9.41	7.33
1.5	8.24	7.08	10.32	8.19
2	10.12	8.16	12.48	10.59
3	N/A	10.32	16.80	16.80
4	N/A	12.48	21.12	21.12

interface not sheared, moderately weak floor and floor/coal interface sheared). Note that the pillar is very long in the direction of entry penetration and the formula only applies at least 50 m from the portal and at least 25 m from the end of the entry (so that end effects are not important):

$$S_{CSIROPART} = 6.36^*(0.41 + 0.59 W/h) \quad [0.5 \leq W/h \leq 2.0] \tag{6.6}$$

where S denotes the pillar strength.

The Bieniawski squat pillar (square cross section width w, height h) formula is:

$$S_{BIENIAWSKI} = 6.00^*(0.64 + 0.36 w/h) \tag{6.7}$$

Wagner (1974) proposed the equation:

$$S_{WAGNER} = 6.00^*(0.64 + 0.36 W_e/h) \tag{6.8}$$

with $W_e = 2W$ for very long rectangular pillars of cross-sectional width W and height h.

It is noted that the slope of Equation (6.6) is much greater than that of Equations (6.7) and (6.8).

By defining $W_e = [(0.69 + 0.44 W/h)W]$ for $0.5 \leq W/h \leq 3.0$ and $W_e = 2W$ for $W/h \geq 3.0$ in Equation (6.8), the proposed CSIRO empirical formula becomes:

$$S_{CSIRO} = 6.00*[0.64 + 0.36(0.69 + 0.44 W/h)W/h] \quad \text{for } 0.5 \leq W/h \leq 3.0 \tag{6.9}$$

and Equation (6.8) for $W/h \geq 3.0$.

Tabling strength values for Equations (6.6), (6.7), (6.8) and (6.9) gives.

Note that Equation (6.9) gives a close match with Equation (6.7) for $0.5 \leq W/h \leq 2.0$ and tends towards Equation (6.8) for $2.0 \leq W/h \leq 3.0$.

The gradual increase in W_e/W in Equation (6.9) from $W_e/W = 0.91$ to Wagner's value of $W_e/W = 2$ (Wagner, 1974), as W/h increases from 0.5 to 3, is demonstrated in Table 6.10, together with an explanation of the effect of confinement on the values.

Figure 6.18 shows the comparison of the four curves. The curve labelled "Average Trendline" is Equation (6.6). Note that the highwall mining pillar analysis was not performed for the value of $W/h = 3$, as this is outside the range of any highwall mining pillar. However, it is the range where Wagner reports his experiments as more reliable, so the curves are assumed to meet there.

Table 6.10 Increase in W_e/W and pillar peak strength for Equation (6.9) as W/h increases.

W/h	W_e/W	Pillar peak strength	Explanation
0.50	0.91	4.92	This pillar will not develop strength beyond the mass UCS value of 4.8 MPa – since the yield criterion is based on σ_1 versus σ_3, confinement in the 3rd direction (σ_2) will not assist strength
0.71	1.00	5.38	This pillar will develop very little strength beyond the mass UCS value
1.00	1.13	6.27	As for 0.5 with a very small effect
1.29	1.25	7.33	Confinement has a small beneficial effect on W_e
1.50	1.34	8.19	Confinement has a small beneficial effect on W_e
2.00	1.56	10.59	Confinement has a beneficial effect on W_e, but not the full effect demonstrated by Wagner
3.00	2.00	16.80	Confinement has a definite effect, demonstrated by Wagner

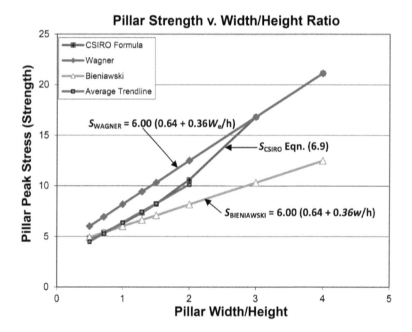

Figure 6.18 Empirical pillar peak strengths by Equations (6.6)–(6.9).

6.7.3 Preliminary estimation of percentage extraction for a given HWM reserve

A preliminary estimation of the extraction ratio for a proposed HWM reserve is offered here. It is emphasised that variability of geological conditions as well as uncertainty in some of the parameters, particularly the mass coal strength, means that it can only give a preliminary estimate, intended to help in the feasibility study for a Highwall Mining Reserve (HMR). If it is decided that highwall mining may be a suitable option for the pit, the geotechnical information will almost certainly have to be refined, in

Table 6.11 The eight key parameters together with example values.

Parameter	Value
Entry width (metres)	3.5
Seam thickness (metres)	3.125
Mining height (metres)	3.125
Average mass cubic coal strength (MPa)	6
Vertical height of highwall (metres)	60
Cover depth at maximum penetration (MPa)	105
Maximum penetration (metres)	300
Desired factor of safety	1.3

Table 6.12 Rating of the roof and floor and interface conditions with example values.

Parameter	Rating
Roof condition	80
Roof/Coal interface condition	80
Floor condition	20
Floor/Coal interface condition	20

order to do a proper layout design. If any features such as jointed coal, jointed roof or very weak floor are present (noting that very weak roof may preclude mining anyway, Section 6.4), they must all be taken into account in the design. The presence of faults or benches or sections of bad highwall, which may imply that entries will have to be skipped, thus leaving barriers, must also be considered.

As stated at the beginning of this section, the empirical approach will only work for panels where all entries are parallel, and mining is over a panel of sufficient length that it can be assumed that full cover stress needs to be carried by the pillars in the centre of the panel. For each value of penetration depth with its corresponding cover stress, a predicted pillar width and extraction ratio is given. This assumes that all entries have penetrated approximately 25 m beyond this penetration depth (in order to preclude end effects) and that there are no barrier pillars. The following summarises the procedure and two examples are included to illustrate it.

Table 6.11 gives the list of input parameters. The example values are those pertaining to Pit 20D Upper at Moura Mine.

It was noted that the loss of confining stress in the pillar plays a major role in pillar strength. As well as the influence of the shape of the pillar, in particular the cross-sectional width to height ratio, the level of confining stress depends on the strength of the roof and floor and also on the presence or absence of sheared interfaces between the coal seam and the roof and floor.

An assessment of the strength of the roof and floor and also of the presence or absence of sheared interfaces are entered as in Table 6.12. Because this is more subjective than the parameter values in Table 6.11, and in line with the approach taken for the HMI, they are entered as ratings. The roof and floor condition ratings are entered independently here, although clearly they should be consistent with the HMI values.

Table 6.13 Rating guidelines for the roof, floor and interface conditions.

Parameter	Excellent (100 ≥ rating > 80)	Good (80 ≥ rating > 60)	Fair (60 ≥ rating > 40)	Poor (40 ≥ rating > 0)
Roof condition	Strong to very strong roof rocks; thickly bedded; joints with wide spacing. No alternation and weathering.	Moderately strong roof rock; medium bedded; joints with moderate spacing. Slight alternation and weathering.	Moderately weak roof rocks; thinly bedded; joints with close spacing. Alternation and weathering.	Weak roof rocks; very thinly bedded; joints with very close spacing. Strong alternation and weathering.
Floor condition	Strong floor rocks. Unaffected by water.	Weak to moderately strong floor rocks. May be wet but not muddy.	Weak floor rocks. Occasionally muddy.	Very weak floor rocks (clay). Wet and muddy throughout.
Roof/Coal or Floor/ Coal interface condition	Sharp cemented interface between coal and strong rock (will often be an erosional contact) with very widely spaced >2,000 mm joints.	Discontinuous horizontal defects along the interface open less than 1 mm.	Discontinuous horizontal defects along the interface open less than 3 mm.	Continuous structural defect associated with very low-strength materials such as slickensides or clay bands.

Table 6.14 Calculated values of pillar width for Pit 20D Upper, Moura Mine.

Penetration depth (m)	Cover stress (MPa)	Pillar width (m)	Percentage extraction (%)
75	1.62	2.28	60.6
150	1.86	2.58	57.6
225	2.21	3.00	53.8
300	2.58	3.45	50.3

A description of the different conditions for the three key parameters is given in Table 6.13.

Based on the information supplied, a pillar width and a corresponding extraction ratio are calculated and shown in Table 6.14. The results are for cover stresses calculated for various penetration distances for one panel in Pit 20D Upper, Moura Mine.

For a parallel layout, if the desired penetration depth is 300 m, the required pillar width is 3.45 m. However, it may be preferable to limit penetration in some entries, and narrow the pillar. For example, when the penetration was limited to 150 m in every fourth entry, a pillar width of 2.5 m was recommended as acceptable in most parts of the pit; see Duncan Fama et al. (1999a).

Example 2 uses the input parameters shown in Table 6.15. These values are close to those used for the back-analysis performed for Pit P, German Creek Mine, which was reported in Section 7 of Report 616F (Shen and Duncan Fama, 1999b). A panel instability occurred during highwall mining of the southern section of Pit P, German

Table 6.15 The eight key parameters for Pit P, German Creek Mine.

Parameter	Value
Entry Width (metres)	3.5
Seam Thickness (metres)	4
Mining Height (metres)	4
Average Mass Cubic Coal Strength (MPa)	4
Vertical Height of Highwall (metres)	40
Cover Depth at Maximum Penetration (MPa)	66
Maximum Penetration (metres)	300
Desired factor of safety	1.3

Table 6.16 Rating of the roof, floor and interface conditions (Pit P, German Creek Mine).

Parameter	Rating
Roof Condition	0
Roof/Coal Interface Condition	0
Floor Condition	0
Floor/Coal Interface Condition	0

Table 6.17 Calculated values of pillar width for Pit P, German Creek Mine.

Penetration Depth (m)	Cover Stress (MPa)	Pillar Width (m)	Pillar Peak Stress (MPa)	Extraction Ratio
At highwall	1.14	2.96	2.49	54.1%
100	1.19	3.12	2.53	52.9%
200	1.41	3.73	2.73	48.4%
300	1.62	4.36	2.92	44.5%

Creek Mine, after 16 entries were mined. The instability led to the burial of a continuous miner and five cars. It was believed that weak floor and a possible concentration of vertical stress beneath the benched highwall led to the failure. The German Creek mass cubic coal seam strength has been taken as 4 MPa.

The roof and floor in Pit P were rated as very weak. Soft and coaly claystone had been found in boreholes close to a failed panel. During mining operations, after excessive water inflows, an incident was reported where the miner sank into a muddy floor. Thus the possibility of very weak, soft floor with very low strength had to be allowed for. Both the roof/coal and floor/coal interfaces were described as sheared and weak. Hence ratings of zero were given here to all four parameters shown in Table 6.16. The results of the calculations are given in Table 6.17.

Figure 6.19 shows a plot of results from examples 1 and 2 compared with the empirical design curve and also the low-strength trendline curve obtained from previous numerical analysis of models with low-strength coal. Example 2 correctly has

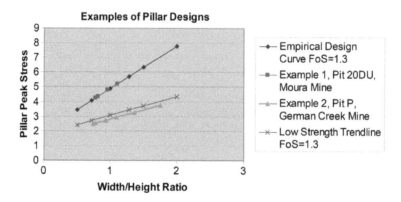

Figure 6.19 Pit 20D Upper, Moura Mine compared with empirical design curve (1) and Pit P, German Creek Mine compared with low-strength trendline, both for FoS = 1.3.

peak strengths about 9% below this curve because of the very low ratings that were assigned for roof and floor parameters.

The mined panel in Pit P consisted of 15 pillars that were formed before failure of the panel occurred. The pillar widths varied irregularly, ranging from a minimum width of 2.1 m to a maximum of 5.4 m, with an average width of 3.67 m. In Duncan Fama et al. (1999c), four different cases for the UCS of the weak floor were modelled. For UCS values between 0.26 MPa and 2.85 MPa the following results were obtained from the modelling. When the vertical stress was 1.34 MPa, failure initiated on the narrower pillars, which had calculated FoS values between 1.15 and 1.30. When the vertical stress was 1.62 MPa, failure initiated on the narrower pillars, which had calculated FoS values between 0.95 and 1.07. The exact FoS for failure of the whole panel could not be calculated, as the calculation could not be continued past this point. For a vertical stress of 1.34 MPa, the empirical calculation here predicted that a regular array of pillars should have a width of 3.66 m to achieve a calculated FoS of 1.34. For a cover stress of 1.62 MPa, the empirical method predicted a pillar width of 3.66 m was required to achieve a calculated FoS of 1.11. Both results are remarkably consistent with the numerical modelling results for the panel for the quoted values of UCS.

It must be emphasised that the empirical calculation cannot cater for very weak, soft floor such as clay. When the floor UCS was taken to be 0.03 MPa, failure of both the narrow pillars and the whole panel was predicted by the numerical model for both vertical stress values above.

A similar verification test was conducted for a highwall mining pit at Ulan Mine. For the coal seam at Ulan Mine, a mass cubic coal seam strength is estimated to be about 7.5 MPa. At a cover depth of 100 m, the required pillar size is 2.75 m. This pillar size agreed with the design and was used during mining.

It is emphasised that the empirical formula may only be used for parallel layout without barriers. For panel layouts with barriers, for example where nine out of ten entries are mined, or six out of seven, or four out of five, it is not possible to give empirical formulae for pillar and barrier widths. However, some estimates of the increases in layout strength may be obtained from examples in Duncan Fama et al. (1999a), and Adhikary and Duncan Fama (2001).

6.8 CASE STUDY

This section presents the back-analysis of one major panel instability at Yarrabee Mine. A highwall mining trial was conducted in Pit B at Yarrabee Mine in late 1997 to assess possible full-scale deployment of highwall mining in the mine. Pit B was a highwall block bounded to the south and west by large displacement faults. A total length of about 550 m was available for mining, including an angled end forming the northern pit boundary; see Figure 6.20. The mining operations in Pit B were conducted by MTA Pty Ltd using a CHM system. A total of 67 entries were mined in this pit in a time span of about six weeks. The penetration distance achieved was on average about 100 m, far shorter than the planned maximum penetration of 350 m. Various problems including roof falls, floor tripping due to soft floor, and gas problems contributed to the poor performance. Despite these problems, the highwall mining panels were observed to be stable during mining.

On the 12th of January, 1999, a localised panel failure occurred in the middle section of the pit (Figure 6.21). The failure extended about 80 m along the highwall and 50 m into the highwall. The failure occurred some 12 months after the completion of highwall mining operations in this pit. The pit gradually filled with water after the pit was closed, and by August 1998 all mined entries were submerged in water.

6.8.1 Pit geology and mining record

Pit B highwall has a SE-NW orientation, and it extends about 455 m in total (excluding the corner section). The height of the highwall is about 50 m. There is a bench about 15 m wide 10 m below the top of the highwall.

The following geological description is given in the highwall mining design report (Highwall Mining Services, 1997). The description is based on two cored bore-holes that were drilled in the northern part of the pit (BHWC37 and BHWC38; see Figure 6.20) and on observations made of the highwall face.

The overburden stratigraphy is typical Rangal Coal Measure sedimentary sequence comprising clayey siltstones, massive to well-bedded, with sub-dominant labile fine grained sandstones and claystone.

Closest to the highwall, the immediate roof comprises a 0.8 m thick claystone, silty, weak to moderately strong, and is gradational with coal at the fused roof contact. Adjacent to the rear boundary fault, a siltstone sequence is observed that extends to the coal roof, replacing claystone. The siltstone is weak to moderately strong rock in which the bedding is reasonably well fused.

The seam comprises dull, to dull and bright-banded compositions in roof and floor plies for a thickness of about 0.8 m. The central seam is bright banded. A weak, carbonaceous mudstone band, about 0.02 m in thickness, occurs about 0.4 m below the seam roof. The band was not considered in the design and is not explicitly modelled in this study.

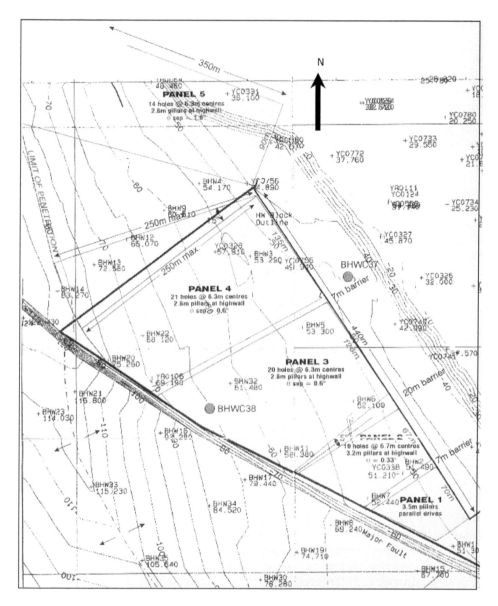

Figure 6.20 Highwall mining panels and design specifications, Pit B, Yarrabee (Highwall Mining Services, 1997).

The average seam thickness is about 4 m. The seam dips to SE with an average dip angle of 8°. The immediate 0.1 m to 0.2 m of floor comprises a weak, but generally intact carbonaceous mudstone interbedded with bright coal. Underlying this thin floor claystone is an interbedded sequence of siltstone and sandstone, which is grey, fine grained and hard. A simplified core log is shown in Figure 6.22, based on the core log from Borehole BHWC37.

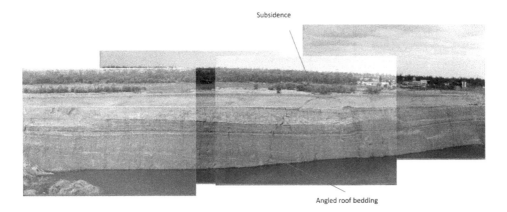

Figure 6.21 Highwall face and a panel failure at the mid section of the highwall, Pit B, Yarrabee. Angled roof bedding planes are noticed at the area of failure.

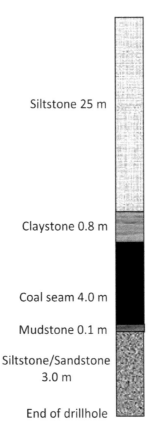

Figure 6.22 A simplified core log of the roof and floor geology in Pit B, Yarrabee.

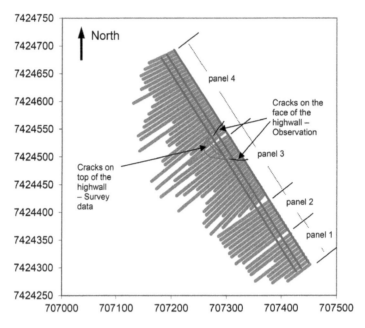

Figure 6.23 Highwall mined entries, and the surface cracks after subsidence, based on survey and observation.

Table 6.18 Designed pillar size and layout in each panel.

Panel No.	Layout	Pillar width at highwall	Entry Nos.
1	Parallel entries	3.5 m	1–10
2	Fanned layout with a radial angle of 0.33°	3.0 m	11–20
3	Fanned layout with a radial angle of 0.58°	2.8 m	21–40
4	Fanned layout with a radial angle of 0.58°	2.8 m	41–67

The most significant structural feature in this pit is the relatively frequently occurring planar slickensided joints observed in the upper to mid plies of the coal seam. The joints dip about 35° to the south along the highwall. The coal joints are persistent within the seam and most extend to the roof and floor. Bedding planes in the roof and floor strata are planar and rough. Bed thickness in the roof is 10 cm to 35 cm.

In the area of failure, roof bedding planes are observed to dip towards the seam at an angle of approximately 15° (Figure 6.21). This feature, however, was not reported in the design report (Highwall Mining Services, 1997).

Highwall mining operations started from the southern end of the pit and progressed to the north. The whole reserve was mined in four panels (Figure 6.23). Panel 5 shown in Figure 6.20 was not mined. The pillar size and layout as designed and used in each panel are listed below (Table 6.18).

Barrier pillars were left between panels. The size of the barrier pillar at the highwall was 7 m, 13 m and 7 m respectively between panels 1, 2, 3 and 4. A total of 67 entries were mined. The penetration depths of the entries are shown in Figure 6.23. The following is a summary of the major events extracted from a review of the mining records:

- The design report recommended that a fixed mining height of 3.1 m be mined and a coal floor of 0.3 to 0.5 m be maintained. A mining height of 2.8 m was targeted during mining to reduce production of high ash coal. Operational problems were encountered in trying to maintain the coal floor. The seam contained numerous small rolls. Because the gamma reading unit was located at the back of the shovel on the continuous miner, there was always a lag between where the cutter head was in the seam relative to roof or floor and the position indicated by the gamma unit. This lack of direct feedback meant that it was very difficult to control the position of the cutting head within the seam. As a result, maintaining constant coal floor thickness had been a major problem throughout the pit.
- When mining operations left 0.5 m of roof coal, roof coal falls would occur. These falls often overloaded the ADDCARs and required numerous withdrawals of the system.
- Cracking was noticed in the front section of one pillar (pillar number was not recorded). When mining operations were some 20 m away from this pillar, the front section of the cracked pillar failed, leading to a localised highwall fall. An investigation revealed that the pillar was supporting a section of highwall hanging outside the original pre-split line.
- While mining entry 43, the deepest five cars (62 m) and the miner were covered by major roof falls. The system was successfully withdrawn with the help of other machinery at Yarrabee.
- Roof falls occurred in subsequent entries after entry 43. As mining advanced the roof falls occurred closer to the highwall. The roof fall problem appeared to have been associated with a section of the reserve where the bedding planes changed from parallel to the seam to dipping down onto the seam.
- Every attempt was made to penetrate beyond the roof fall but without success. Numerous withdrawals and re-entries were made to remove the roof fall material.
- Just as the problems of roof falls seemed to have disappeared at the highwall, a structure was encountered about seven cars in, which persisted in several entries.
- As penetration depth started to increase, methane was encountered and the penetration depths were limited to 6–12 cars (74–148 m). Mining continued to be stopped by methane. Highwall mining operations at Yarrabee were rapidly drawing to a premature conclusion.

It should be emphasised that major roof falls occurred in entries from 43 onward. The subsided area appears to be right behind the area where major roof falls occurred during mining.

6.8.2   Back-analysis of the observed panel failure

The subsidence extended about 80 m along the base of the highwall between entries 33–43 and 50 m into the highwall (Figure 6.23). The area covered nearly half of panel 3, three entries in panel 4, and a barrier pillar between the two panels. The barrier pillar had an initial width of 7 m at the highwall but narrowed to 3.9 m at a penetration depth of 50 m due to the radial entry pattern in both panels 3 and 4.

It is surprising that the subsidence occurred right above a barrier pillar, although this barrier pillar narrowed to close to the normal pillar width at about 50 m inbye. A deterioration in local geological conditions is likely to have contributed to the failure.

The mining records indicate that roof falls were experienced starting from entry 43, which is about 15 m north of the barrier pillar. The operators believed the falls were caused by the change of bedding orientation from parallel with the seam to dipping onto the seam. The change of bedding orientation is clearly visible in Figure 6.21, and it appears that a major bedding plane dips to the seam at the centre of the subsided area. The cores from borehole BHWC37 close to the barrier pillar (Figure 6.20) suggested the roof bedding planes were sheared (Highwall Mining Services, 1997).

A number of factors could have contributed to the failure and were investigated in this study:

• Extensive roof falls caused by the unfavourable bedding, flooding of the pit and deterioration of the strength with time;
• Stress concentration in the vicinity of the highwall;
• Lower coal strength than assumed in design;
• Pillar strength reduction due to time and water infill.

Numerical analysis was conducted using the 2D distinct element code *UDEC* (Itasca, 1996). A number of numerical models were used in this study to investigate the stresses under the highwall, span stability, and the strength of pillars of different widths. In all the models, the simplified geology shown in Figure 6.22 was used, except that the thin mudstone floor was modelled as a weak interface.

Seven core samples from Pit B had been tested for the uniaxial compressive strength (UCS) and elastic modulus (Highwall Mining Services, 1997). Based on this, the mass coal strength for the Yarrabee seam would be estimated by the CSIRO to be 3.0 MPa. This is compared with a mass UCS of 3.6 MPa used in the design (Highwall Mining Services, 1997).

Note that a mass coal UCS is the uniaxial compressive strength of a cylindrical coal mass. It differs from the cubic mass coal strength, which is more widely used in literature. It has also been reported that the mass coal UCS is typically 20-30% less than the cubic strength (Townsend et al., 1977).

The extensive and inclined coal joints in the seam may have had a significant effect on pillar strength. They were considered in the original design and have been considered in this study as well.

The strength and elastic properties of roof and floor rocks have not been tested, nor were any interface or joint properties available. For the purpose of numerical modelling, these material properties have been estimated mainly based on geological descriptions and our previous modelling experience. The material properties used in this study are listed in Table 6.19.

No stress measurement results from Yarrabee Mine were available. In this study, it was assumed that the ratio of principal horizontal stress to vertical stress is 0.33 in the direction parallel to the highwall, and 1.0 perpendicular to the highwall, as it is

Table 6.19 Geomechanical input parameters for numerical modelling.

	E (GPa)	ν	Cohesion (MPa)	ϕ (°)	σ_t (MPa)	Mass UCS (MPa)
Coal	2.0	0.37	Hoek-Brown $m = 2.93$, $s = 0.075$, $\sigma_c = 16.2$ MPa, $a = 0.65$			3.0
Siltstone	4.0	0.25	1.3	35	0.5	5.0
Claystone	2.5	0.25	1.04	35	0.4	4.0
Coal/roof/floor interfaces			0.012	16	0	
Other interfaces			0.5	30	0	
Coal joints			0.1	33	0	
Rock joints			0	30	0	

close to the normal direction of the regional fault. These stress ratios were assumed to be the same in both coal and other rocks, prior to the excavation of the highwall.

Previous studies (Kelly et al., 1998; Duncan Fama et al., 1999a) suggest that stress concentration could occur near the highwall face due to the effect of an overlying bench, and could lead to elevated stress in the coal seam beneath the bench. The possibility of high stress in Pit B at Yarrabee Mine was investigated. A 2D numerical model simulated a cross section perpendicular to the highwall prior to highwall mining extraction. The highwall profile in the area of subsidence was used. In this model, intact rocks and coal were modelled as elastic materials and their properties are listed in Table 6.19. All interfaces and joints except for the roof and floor interface were assumed to have a friction angle of 30° and cohesion of 0.5 MPa. The roof and floor interface was assigned a friction angle of 16° and cohesion of 0.012 MPa. The Pit B highwall appears to be smooth and free from major fractures. Therefore, no blast damage was included in the model.

As mentioned before, the *in situ* horizontal stress perpendicular to the highwall was assumed to equal the vertical stress before highwall excavation. The stress model is shown in Figure 6.24. The predicted vertical stress at the mid-height of the seam is plotted in Figure 6.25. The maximum principal stress is also plotted in the figure. At the toe of the highwall, a vertical stress of 0.87 MPa and a maximum principal stress of 1.05 MPa are predicted. The maximum principal stress is predicted to be higher than the vertical stress within a penetration distance of 70 m. The vertical stress appears to be increasing gradually with the penetration distance, whereas the maximum principal stress reaches a local peak of 1.5 MPa at 40 m, where the vertical stress is 1.25 MPa. At penetration depths greater than 70 m, the two stresses are generally the same, indicating that the disturbance to stress field due to the highwall excavation diminishes away from the highwall.

The failure criterion used in the coal strength model depends on the disparity between the maximum and minimum principal stresses, not the difference between the vertical and horizontal stresses. In order to study the pillar stability, a 2D vertical cross section parallel to the highwall (perpendicular to the 2D plane shown in Figure 6.24) is often used in practice, and the vertical stress in this cross section is often assumed to be the maximum principal stress. In our case, because the maximum principal stress

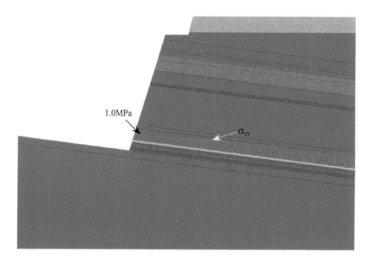

Figure 6.24 A 2D model to study the stresses near the highwall. The vertical stress is plotted.

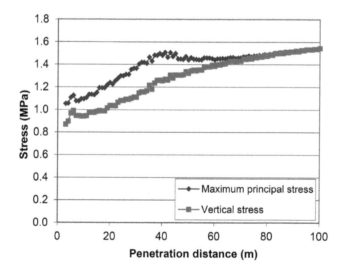

Figure 6.25 Numerically predicted vertical stress and maximum principal stress in the coal seam in the vicinity of the highwall.

is not vertical and nor is it in the 2D vertical plane, a 3D study seems to be the best approach to give an accurate prediction of panel strength. However, due to limitation in model size and the number of elements, a coarse 3D model may produce no better results than a more detailed 2D model. For simplicity, a 2D approach was adopted in the following sections to study the pillar/panel stability.

Because the maximum principal stress is not in the 2D vertical cross section, its effects on the results obtained by 2D study are uncertain. To consider the unknown

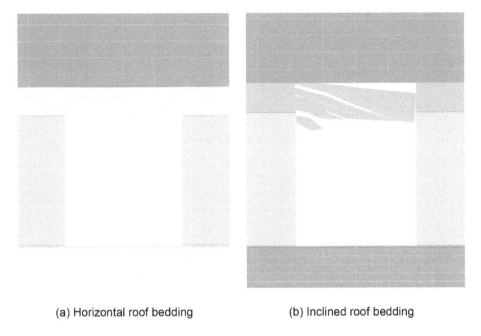

(a) Horizontal roof bedding   (b) Inclined roof bedding

Figure 6.26 Stable and unstable spans for horizontal and angle roof bedding, numerical predictions.

effect from the out-of-plane principal stress on the results of a 2D analysis, both the vertical stress and the maximum principal stress as obtained in this 2D section were used.

The stability of a highwall mining entry span at Yarrabee was investigated using a *UDEC* model. Two cases were studied and compared: (1) when the bedding partings in the roof claystone were horizontal and (2) when the roof bedding partings dipped at 15° towards the seam. In both cases, the full seam height was assumed to have been mined so that the bedded claystone is exposed as the immediate roof. Vertical joints in the claystone layer were modelled at a spacing of 3.5 m.

Figure 6.26 shows a comparison of modelled roof stability for the two cases. When the bedding planes are parallel to the seam, the layers act as fixed-end or simply supported beams after delamination. The beams under mainly their self-weight are predicted to be stable, and hence the span is predicted to be stable. When the bedding dips towards the seam at an angle, the layers behave as cantilever beams, and in this case the span is predicted to be unstable.

Roof falls are expected to be more severe at a shallow penetration depth near the highwall. This is because (1) the rock mass near the highwall often suffers disturbance, and (2) the confinement stress in the direction perpendicular to the highwall is low. No evidence of more frequent roof falls close to the highwall has been recorded in this pit during mining. However, most of the roof falls are expected to occur after mining, particularly after the entries have filled with water.

Table 6.20 Pillar strengths for various conditions, predicted by numerical modelling.

Pillar width (m)	Pillar strength (MPa)		
	Case 1 (height = 2.8 m)	Case 2 (height = 4.8 m)	Case 3 (height = 4.8 m, weaker coal joints)
2.8	3.2	3	2.2
3.2			2.2
3.7	4.8	3	2.6
4.4	6.1	4.2	3.3
7.0		5.2	4.1

Within the subsided region (Figure 6.23), there were both the regular pillars and a barrier pillar. The barrier pillar had a width of 7 m at the highwall and narrowed to 3.9 m at a penetration depth of 50 m, based on the design. The regular pillars ranged from 2.8 m wide at the highwall to 3.3 m wide at a penetration depth of 50 m.

The subsidence was apparently caused by the failure of 12 pillars, which had widths ranging from 2.8 m to 7 m. To analyse the instability, it was necessary to investigate the strength of a pillar of various widths.

Taking advantage of the boundary symmetry, an infinite row of pillars was effectively modelled with only a pillar and a half of an entry on either side of it. The mechanical properties listed in Table 6.19 were used as the base case for this study. The following variations from the base case were investigated:

- Case 1: Pillar/entry height = 2.8 m, and coal layer of 0.6 m was left in the roof and floor. This is an ideal case configuration which the mining operators tried to achieve. It is uncertain how well this was achieved during actual mining. Coal falls and cutting into the high ash floor were reported in the mining record.
- Case 2: Pillar/entry height = 4.8 m. An entry this high could have resulted from mining too close to the floor combined with extensive falls of roof coal and roof claystone. This was likely in the region of subsidence due to the angled roof bedding. Mining records reported regular coal falls during mining in this region. The claystone roof with inclined bedding was unstable once exposed. Time and water infilling in the entries would have worsened the roof falls.
- Case 3: Pillar/entry height = 4.8 m, plus weaker coal joints. It is considered that this case represents the most likely scenario. A cohesion of 0.1 MPa and friction angle of 33° were estimated in the design report for the inclined coal joints through the seam. This estimate may reflect the *in situ* strength conditions. More than one year after mining, particularly after the entries were flooded, the coal joints would be expected to have lost their cohesion and their friction angle might have been reduced. A zero cohesion value and a friction angle of 30° were used in case 3.

A number of pillar sizes were modelled for all three cases. The results are summarised in Table 6.20 and plotted in Figure 6.27.

As expected, the pillar strength in case 1 is the highest among all three cases. When the pillar height increases from 2.8 m for case 1 to 4.8 m in cases 2 and 3, the pillar strength is found to reduce significantly, particularly when the pillar width is over 3 m.

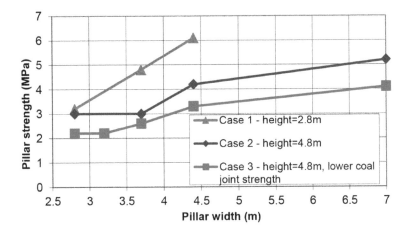

Figure 6.27 Variations of pillar strength with pillar width for three different cases.

In case 2, a pillar strength of 3 MPa is obtained for pillars ≤3.7 m wide. This strength is equal to the mass coal UCS used in the modelling. This result suggests that the effect of the coal joints with the particular strengths that were assigned has only a minor effect on pillar strength.

In case 3, which was formulated to simulate the effects of time and water, the pillar strength is found to reduce by up to 30%. This reduction is significant and based on purely theoretical considerations requires a minimum FoS of 1.4 at design to achieve a medium-term pillar stability.

Having evaluated individual pillar strength under various mining scenarios, the next phase of the investigation was to evaluate overall panel strength. To do this, it is necessary to sum the strengths of the individual pillars, multiplied by their widths, and divide that value by the overburden stress.

The average strength of the panel in the subsided area can be calculated from the individual pillar strengths obtained in case 3. The average panel strength is calculated in a number of 2D vertical sections parallel to the highwall and at various penetration depths, using the following formula:

$$Average\ Panel\ Strength = \frac{\sum (Pillar\ Width \times Pillar\ Strength)}{Panel\ Width} \tag{6.10}$$

After calculating the average panel strength, the FoS at a specific penetration depth can be then obtained by using the loading stress obtained earlier. Here both the vertical stress and the maximum principal stress have been used.

$$FoS = \frac{Average\ Panel\ Strength}{Actual\ Stress} \tag{6.11}$$

The results are summarised in Table 6.21 and plotted in Figure 6.28.

In Table 6.21, ranges of actual stress values are given that were used in the calculation of FoS. The upper bound of the range is the maximum principal stress and the

Table 6.21 Calculated average panel strength and FoS in the subsided area.

Penetration (m)	In-panel pillar width (m)	Barrier pillar width (m)	Failed panel width (m)	Average panel strength (MPa)	Actual stress (MPa) (σ_v, σ_1)	Factor of Safety (FoS)
0.00	Penetration depth less than 10 m is excluded in this calculation because the plane					
5.00	strain assumption used in the study is no longer valid.					
10.00	2.90	6.39	70.0	1.19	0.94–1.13	**1.06–1.27**
15.00	2.95	6.08	63.7	1.20	0.99–1.15	**1.01–1.22**
20.00	3.00	5.78	57.4	1.22	1.03–1.29	**0.98–1.18**
25.00	3.05	5.47	51.0	1.24	1.09–1.31	**0.94–1.13**
30.00	3.10	5.17	44.4	1.26	1.16–1.42	**0.88–1.08**
35.00	3.15	4.86	37.8	1.28	1.24–1.48	**0.87–1.03**
40.00	3.20	4.55	31.1	1.32	1.26–1.51	**0.88–1.05**
45.00	3.26	4.25	24.3	1.38	1.31–1.45	**0.95–1.06**
50.00	3.31	3.94	17.4	1.45	1.35–1.45	**1.00–1.08**
55.00	3.36	3.64	10.4	1.57	1.38–1.46	**1.08–1.14**

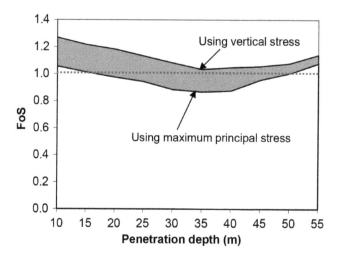

Figure 6.28 Variations of Factor of Safety (FoS) with penetration depth in the subsided area.

lower bound is the vertical stress. The values were taken from the numerical modelling results. As a result, a range of FoS was obtained and is given in Table 6.21. The true FoS should lie within the given range.

Figure 6.28 presents the FoS variation with penetration distance for the failed panel for case 3 conditions. It shows that the FoS reaches a minimum at a penetration distance of 35 m, and the minimum FoS is predicted to be in the range 0.87–1.03. At a penetration distance of 15–50 m, if the maximum principal stress is used, the FoS is predicted to have fallen below 1.0.

Judging from the plot of FoS presented in Figure 6.28, it is believed that the subsidence started at about 30–40 m inbye, which is just behind the top bench of the highwall. The failure propagated both towards the highwall and deeper into the

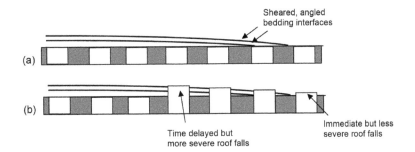

Figure 6.29 Roof falls of different sizes at different locations due to an angle roof bedding.

panel. However, due to stronger confinement deeper in the reserve, the extent of the subsidence was further towards the highwall than into the panel.

During the above analyses, a pillar height of 4.8 m (case 3) was used. It is believed that the maximum increase in height of the pillars occurred only in the subsided area due to extensive roof falls. In the subsided area, the steeply dipping overburden caused the roof mechanical behaviour to depart most from that of a beam held in compression. The mining records reported that the worst roof falls experienced during mining occurred in entry 43, towards the northern end of the subsidence zone. The locations of roof falls experienced during mining may reflect the intersection between the angled roof layers and the seam. It is possible that more severe roof falls may have occurred in the area behind the intersection after mining, given that the immediate roof is sheared. It is postulated that some form of progressive failure took place. This mechanism is demonstrated in Figure 6.29. In entries 33–42 where the subsidence occurred, although no major roof falls were reported during mining, it is possible that extensive roof falls occurred after mining, which eventually caused the failure.

Previous experience has shown that highwall mining panel failures can occur rapidly, sometimes violently, and often extend widely through the mined panel. The subsidence at Pit B showed the opposite characteristics. It occurred rather slowly (the full subsidence developed over two weeks) and was confined to a limited area. The wide barrier pillar (7.0 m) is believed to be the key factor for the slow panel failure.

It is known that a wide "fat" barrier pillar behaves quite differently from a narrow and "skinny" pillar. While a skinny pillar often collapses quickly, a fat pillar may fail progressively. This is because the confinement from the roof and floor contact often restricts the volume expansion associated with coal failure, and hence increases the confining stress in the centre of the wide pillar.

It is believed that the failure of the 7.0 m barrier pillar in Pit B at Yarrabee occurred progressively, which in turn led to a progressive panel failure as observed.

6.8.3 Conclusions

The back-analysis results presented above suggested that the subsidence in Pit B might have been caused by three major factors:

- Angled roof bedding over a limited area led to severe roof falls, which heightened the pillars and reduced the pillar strength;

- Water infill into the mined entries contributed to the roof falls and weakened the pillars, mainly by weakening the coal joints;
- Time effects also led to progressive roof falls and pillar weakening, as observed in other highwall failures.

Several lessons should be learnt from the failure at Yarrabee:

- A better understanding of the possible effects of geological variation across the reserve is necessary;
- Measures should be taken to prevent flooding of the highwall mined entries if long-term stability is a major concern;
- A higher FoS is required to accommodate the effects of time on overall pillar/panel strength. However, at present the effects of pillar deterioration with time are not able to be quantified.

6.9 MONITORING OF HIGHWALL MINING OPERATIONS

A highwall mining reserve typically extends over an area of 0.3 to 0.5 km^2. A range of geological and geotechnical conditions will be encountered over an area of this magnitude, particularly in a variable sedimentary environment. Consequently, it is inevitable that in localised areas geotechnical conditions will vary from the design assumptions to such an extent as to render the design invalid and possibly create mining instabilities. Although a good design will leave sufficient factor of safety to cater for a certain amount of variability, it is important that the mining conditions are constantly monitored to check for the unforeseen variations. If any major alteration in mining conditions occurs, or there is any forewarning of major instability, remedial measures must be taken to protect mining personnel and equipment and to ensure a smooth and continuous mining operation.

Monitoring of the mining conditions can be divided into two types:

- Visual monitoring by the mining crew;
- Instrumental monitoring of the mining panel.

6.9.1 Visual monitoring of mining conditions

Highwall mining systems are equipped with tools that can also perform exploration functions. A highwall miner has cameras installed at the cutter head, which are used to observe the cutting process and mining conditions. The relative hardness of the coal and rock can be easily assessed by recording the electrical current flowing through the cutter head when the particular material is being cut. The condition of the coal seam and possible roof falls can be examined by checking the material on the conveyer belt as it arrives at the portal. All these observations are routinely carried out as part of the mining operations. The observations provide invaluable first-hand information on the local mining conditions.

A number of geotechnical parameters affect mine stability and these have been identified by Duncan Fama et al. (1999a). It is recommended that these parameters be given special attention during visual monitoring. They are listed below.

- Roof conditions, including roof falls, roof strength, roof jointing etc. Any roof falls that occur immediately behind the continuous miner can be observed through the camera and should be recorded. Roof falls in other parts of the entry may not be observed by the camera but can be identified by rocks on the conveyer belt and should also be recorded. Changes in roof hardness during cutting, sudden increases in the intensity of jointing and deterioration of roof conditions should all be monitored.

- Pillar conditions. Any observed pillar spalling should be reported immediately. Pillar spalling could be a sign of an overstressed pillar and imminent pillar failure. However, it could also be caused by localised coal falls due to the existence of coal joints. The cause of pillar spalling should be determined by an experienced mining or geotechnical engineer.

- Floor conditions. Soft floor may significantly reduce the pillar/panel strength, due to the limited floor bearing capacity or slipping along the floor interface. Soft floor is noticeable during mining, particularly in wet ground. It is often identified by a record of frequent tripping of the miner or related mining difficulties in the entry. Extensive areas of soft floor should be reported to the geotechnical personnel and alteration of mining layout may be needed.

- Major geological structures. Many operational difficulties (roof falls, floor tripping, water etc.) occur in the vicinity of geological structures. Depending on the nature and the extent of the geological structure, the mining difficulties may be localised areas, or they may lead to large-scale pillar/panel failures. The structures found during mining should be recorded and reported. Where possible, they should be examined by the geotechnical personnel to assess their effect on the layout stability both for the current design and future designs.

- Changing mining configuration. The mining height may be changed from the design during mining, which consequently changes the pillar height. Roof coal may be left during mining. An individual pillar size may be altered in response to operational requirements. All these changes will affect the factor of safety of the mining layout. Any changes from the design need to be recorded and their effects on the pillar/panel stability need to be evaluated.

Any significant differences between the design and its underlying assumptions and how the design is implemented should be fed back to the design team. The impact of the alteration in mining conditions on the mining layout stability needs to be evaluated and if necessary a new layout should be designed.

The following example demonstrates the importance of this monitoring and feedback process.

In November 1997 a continuous highwall mining system in operation in a pit mining the B lower seam at Moura Mine became trapped in the 21st mined entry as a result of a massive instability. Fallen material trapped the continuous miner, the lead car and 11 conveyer cars. The layout design for the pit had followed the guidelines which had been successfully applied to nine previous pits at Moura Mine. The average seam thickness in the pit was 4.8 m with a mining height of 3.7 m. Mining to the base of the seam left a coal layer typically 1.1 m thick in the roof of each entry, which was predicted to be stable. During initial operations, the design appeared adequate, although roof coal falls were reported during mining. The most extensive falls were recorded in entry 7. There was also anecdotal evidence of falls in other entries subsequent to their being completed.

The extensive roof falls effectively increased the pillar height from 3.7 m, as designed, to an actual 4.8 m. A comprehensive back-analysis conducted after the failure (Kelly et al., 1998; Duncan Fama et al., 1999b) indicated that, when the pillar height increased from 3.7 m to 4.8 m, its calculated factor of safety reduced from 1.3 to 1.1 at an overburden stress of 2.4 MPa, as was used in the design. If the stress concentration in the vicinity of the benched highwall is also taken into account, then the calculated factor of safety is well below 1.0, so that pillar failure was predicted. This is particularly true when the mining panel exceeded the critical panel width (21 entries and a total panel width of about 150 m), and the stress redistribution to the abutment was no longer sufficient to accommodate the excessive stress in the panel.

The significant change in the pillar height required a redesign of the pillar width and mining layout to ensure the stability of the panel. Unfortunately, the impact of the extensive roof falls on the panel stability was not realised by the mining operators, and the new mining information did not feed back to the design team. As a result, a major panel failure occurred.

6.9.2 Instrumental monitoring of the mining panel

Two geotechnical monitoring tools have been tested by CSIRO for their suitability to monitor pillar/panel instability. They are:

- Deep-hole surface extensometers
- Microseismic monitoring techniques

In a field trial both the monitoring systems were installed in a mining panel before mining was carried out in Pit 20 DU, Moura Mine. The response of the panel to highwall mining was monitored during and after mining. The extensometers were installed in the roof spans of two entries to monitor the displacement of the spans. The microseismic monitoring geophones were installed over the mining panel so that seismic events could be detected and recorded. Details of this monitoring are given in Shen and Duncan Fama (1999b).

The monitored pit was stable, and hence the characteristics of critical roof displacement or microseismic events for panel failure were not obtained through this monitoring. The microseismic activity that was recorded is believed to have arisen from stress change in the rock mass and effect of major geological structure. However, by comparing the microseismic intensity for the stable panel in Pit 20DU with that for a longwall panel (caving), the characteristics of microseismic events in the transition zone from stable to unstable have now been better defined.

Figure 6.30 shows the difference in microseismic intensity for the stable panel at Moura and a longwall panel at Southern Colliery.

The seismic events recorded during mining of Pit 20DU at Moura were at least two orders of magnitude weaker than the events detected at the southern longwall, which were associated with roof collapse processes. As no abnormal stability problems were encountered at Pit 20DU during the microseismic monitoring period, the events with energy release levels from 10^{-5} to 10^{-1} (m/s)^2 may imply that a catastrophic roof

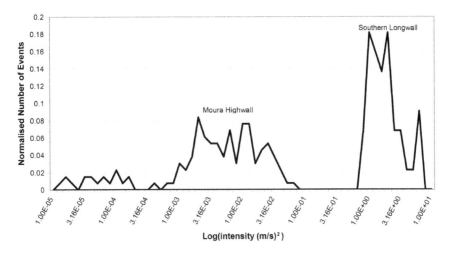

Figure 6.30   Seismic intensities detected at Pit 20DU at Moura Mine and a longwall panel at Southern
Colliery.

problem is unlikely. However, if the energy level reaches 1 to 10 $(m/s)^2$ the possibility
of impending pit stability problems should be seriously considered as this energy level
has been associated with significant rock failures underground.

The maximum observable distance for the seismic events recorded at Pit 20DU
was calculated to be about 200 m. For stronger events such as would be associated
with roof failures, the detection distance is likely to be more than 250 m.

The deep-hole extensometers only recorded limited elastic displacement in the
stable panel of Pit 20DU. An increasing displacement in the overburden roof may
indicate that progressive caving is occurring and actions should be taken to limit the
risk of panel failure.

Further site tests of the two panel monitoring techniques are required to identify
the precursor signals for a catastrophic panel collapse. Other monitoring innovations,
such as highwall photogrammetry and radar monitoring, may also be suitable for
monitoring panel behaviour.

6.9.3   Conclusions and recommendations

Two types of monitoring of the mining conditions have been discussed above, i.e.

- Visual monitoring by the mining crew;
- Instrumental monitoring of the mining panel.

Visual monitoring is easy to do and often bears no additional cost (except it may
slow the mining operation slightly). It will also enhance communication and feedback
and thus will be a potential bonus to mine operators. Visual monitoring could and
should be conducted routinely regardless of the scale of the operation.

Instrumental monitoring is more accurate and can provide the mining operators with information that cannot otherwise be obtained by visual monitoring, such as the panel displacement and pillar instability. However, instrumental monitoring is relatively expensive.

It is hard to quantify in what situation instrumental monitoring should be used, since it is obviously related to the overall economic feasibility for the pit. It is however recommended that, where a panel instability is too costly (sterilising cost resources and losing CHM equipment) or is environmentally intolerable, instrumental monitoring should be considered. The cost of monitoring against that of a catastrophic panel failure is often small.

6.10 SUMMARY

Successful highwall mining requires successful planning, design and mining. This section provides guidelines and tools for each of the three stages. They are summarised below.

6.10.1 Mine planning

A potential highwall mining pit needs to be evaluated for its technical and economic feasibility. A number of minimum conditions must be satisfied for highwall mining operations to even be a possibility, such as poor roof conditions, minimum seam thickness, minimum pit width, etc. If any of these minimum conditions are not met, the given pit will not qualify for highwall mining. When these minimum conditions have been met, a detailed feasibility evaluation is required. The Highwall Mining Index (HMI) presented in the report provides a practical tool for this purpose. The HMI is an overall rating of pit conditions for highwall mining, and it considers 15 important parameters such as roof condition, highwall condition, seam condition etc. A HMI greater than 60 indicates a potentially good site for highwall mining, whereas a HMI less than 40 indicates the site is unlikely to yield a profitable highwall mining operation.

Accurate assessment of pit mineability using the HMI requires knowledge of geological and geotechnical information. The extent of this knowledge will determine the confidence level that can be assigned to any HMI determination. A HMI with high confidence level indicates the site is well understood and the HMI is reasonably reliable. In contrast, a HMI with low confidence level implies that the assessment results are doubtful and more information is needed for a more accurate assessment.

A similar index system was developed for auger mining, the Auger Mining Index (AMI). The AMI is mainly based on the HMI, but some parameters in the HMI (such as roof and floor conditions, mining reserve, stone band etc.) are re-evaluated and/or given a new rating based on their effect on the auger mining.

6.10.2 Site investigation and layout design

Once a pit is considered to be potentially suitable for highwall mining, a detailed site investigation and mine design should be performed. The site investigation should obtain detailed geological and geotechnical data in the mining reserve. It will involve all or any of: drilling additional boreholes in the mining reserve, performing geophysical logging, testing of carefully sampled rock cores in the laboratory and geological modelling of the reserve.

The new geological and geotechnical data are primarily required for the detailed layout design. However, if the new information is substantially different from that used in evaluating the HMI then it may be appropriate to re-run the HMI analysis. It is possible that the initial conclusion about the mineability of the pit may need to be changed based on the new information.

The key components in mine design include pillar and layout design, panel stability assessment and span stability assessment. Pillar and layout design include specifications for pillar size (height and width) and layout arrangement (parallel layout or radial layout, barrier pillar arrangements etc.) and require adequate geological/geotechnical data to be used in association with the empirical and numerical tools.

The designed layout should meet the stability requirement both for the individual pillars and the panel as a whole. Pillar stability is managed by specifying a minimum calculated Factor of Safety (FoS). The value of FoS that is adopted for a mine design will depend on the importance of maintaining a stable panel, the level of confidence in the input data and the methods used in the analysis. If the guidelines and design tools provided in this report are followed, a calculated FoS of 1.4 is recommended for initial designs in pits where long-term subsidence is not a concern. Once mining experience has been obtained in the pit or in similar conditions in nearby pits, the mine management may choose to lower the FoS to 1.3. If a pit needs to be permanently stable due to environmental or engineering constraint, a higher FoS should be used; however, the lack of long-term monitoring data means that it is not possible to make any general recommendations regarding appropriate values.

The layout should also be checked for overall panel stability in order to minimise the risk of major catastrophic panel failure. The panel stability evaluation method given in Adhikary et al. (2002) can be used for this purpose. If the layout does not meet the criterion for panel stability, preventative measures should be taken, such as leaving barrier pillars at suitable intervals.

Span stability should be assessed using the tools described in Section 6.6. Depending on the assessment results, remedial measures may need to be taken in order to achieve as smooth and as successful a highwall mining operation as possible. These measures may include leaving roof coal to reduce the incidence of roof falls, or using an alternative mining system (e.g. auger) that is more tolerant of adverse roof conditions.

6.10.3   Mining operations, monitoring and feedback

If the mine layout departs from the design, for whatever reason, the design team needs to be consulted in order to assess the risk arising from those changes.

Actual geological and geotechnical conditions often depart from those used in the design. Visual monitoring of mining conditions during operations is important as it provides detailed information on the varying geological and geotechnical conditions for virtually no additional cost. Any substantial change in mining conditions, such as extensive roof falls, pillar spalling, encountering major faults etc., must be reported to the geotechnical personnel and the effect of such a change on the roof/pillar/panel stability should be evaluated.

In critical situations, where there is no prior performance history, instrumental monitoring may be needed. Deep-hole surface extensometers can be used to monitor localised movement of the roof rock during mining. If located in the vicinity of the initiation point of a panel failure, extensometer data could also be used to forecast an upcoming panel failure, if the displacements start to increase sharply. Microseismic monitoring provides an alternative but more expensive tool that can be used to monitor a greater volume of the mining panel. The characteristics and the location of the microseismic events can be used to analyse the stability of the pillars and the panel. Results have only been obtained from both these techniques monitoring a stable highwall mining panel. Future research should be directed to obtaining data from monitoring a highwall mining panel as it fails.

Chapter 7

Highwall mining hazards and mitigation

7.1 INTRODUCTION

The understanding of the stability of structures within highwalls is critical for a productive and safe working environment. In most mines, geotechnical issues are considered a high priority and designated 'principal hazards' and are part of the Principal Hazard Management Plans (PHMPs). Besides the structural instability of highwalls, the safe operation of the Continuous Highwall Miner (CHM) inside the highwall entries, from possible roof collapse, percentage of inflammable gas in the general body of air, stagnation of water inside the entries and nearby blasting operations, are some of the hazardous issues that need to be addressed properly for the safe and productive operation of highwall mining. Most of these issues have already been discussed separately in Chapters 4 and 6 in different case examples. It is necessary to address such issues during the mine design process in order to reduce the likelihood of stability and safety issues within the constraints of the project budget and time limit. Detailed discussion of these processes is contained in *Guidelines for Open Pit Slope Design* (Read and Stacey, 2009). Geologic features, such as faults, bedding planes and joints control, strongly influence potential highwall failure modes. Attention has now been focussed on the fly ash backfilling of highwall entries, which is an environmentally friendly solution for the stability of web and barrier pillars with better recovery of coal in abandoned mine remediation. Special consideration is also needed for folded geology. The risks associated with this have been a concern in highwall mining for smooth operation under specific rock-geologic conditions.

7.1.1 Incidents statistics of highwall mining

The Mines Safety and Health Administration (MSHA) reviewed 20 years' worth of records of incidents in highwall mining and augur mining from 1983 to 2002, which included 5,289 highwall mining holes (Zipf and Bhatt, 2004). They could identify nine fatalities in total, of which ground control issues contributed to three (1/3rd of the total fatalities). From the frequency of incidents it was concluded that its accident proneness was on a par with surface coal mining, and hence could be considered as a very safe modern mining method. Among the ground control issues, it was also found that slope stability was the major contributor for serious accidents causing fatalities. The reportable incidents identified by the MSHA during the above 20-year period included 605 incidents at auger and highwall mining operations. These were categorised as

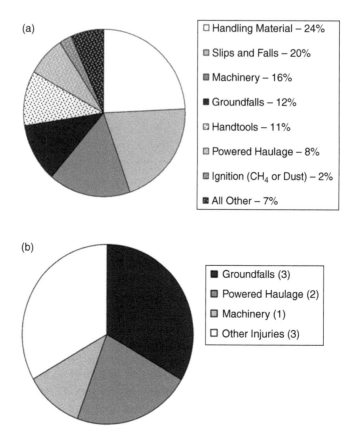

(a)

- □ Handling Material – 24%
- ▣ Slips and Falls – 20%
- ▣ Machinery – 16%
- ■ Groundfalls – 12%
- ▣ Handtools – 11%
- ▣ Powered Haulage – 8%
- ▣ Ignition (CH_4 or Dust) – 2%
- ▣ All Other – 7%

(b)

- ■ Groundfalls (3)
- ▣ Powered Haulage (2)
- ▣ Machinery (1)
- □ Other Injuries (3)

Figure 7.1 Distribution of (a) 605 incidents and (b) fatality causes in highwall and auger mining (Zipf and Bhatt, 2004).

9 fatalities, 460 non-fatal lost-time injuries and 136 non-lost-time injuries. Figure 7.1 shows the distribution of the causes of all the incidents. Out of the 605 incidents recorded, 12% were caused by geotechnical issues such as ground falls. This chapter is focussed on the geotechnical hazards in highwall mining and their possible mitigation measures.

7.2 GEOTECHNICAL HAZARDS

There have been few instances of wall failure during highwall mining. As such, the understanding of slope hazards for coal mining is therefore reliant on observation, experience and judgement. Geotechnical specialists are unlikely to be exposed to slope hazards to the same extent and with the same frequency as coal mine workers. Slope design is an engineering-level risk control mechanism, intended to identify consequences and provide reliable information on likelihoods. Risk management methodology necessitates constant attention to check that control measures are in practice as intended and to identify changed conditions where controls need to be

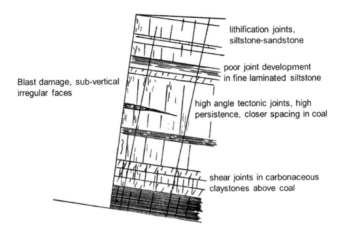

Blast damage, sub-vertical irregular faces

lithification joints, siltstone-sandstone

poor joint development in fine laminated siltstone

high angle tectonic joints, high persistence, closer spacing in coal

shear joints in carbonaceous claystones above coal

Figure 7.2 Typical structures in a highwall (Simmons, 2003).

revised (Simmons, 2013). Nonetheless, a list of ground-related hazards can be drawn up as follows (Simmons, 1995; Simmons and Simpson, 2007):

- Excavated pit wall or batter collapse;
- Dumped spoil slope or batter, or stockpile collapse;
- Release of isolated rock-falls from slopes;
- Heave of roadways or working benches;
- Collapse of cavities under roadways or working benches; and
- Uncontrolled release of water or liquid mud.

For coal highwalls, there are particular geometries and geology and therefore stability issues are different from those generally associated with open-cut mines and hard-rock mining (Seegmiller, 2000). Geologic features such as faults, bedding and joints strongly influence or control the potential highwall failure modes. Special consideration is also needed for folded geology. The risks associated with these structurally controlled failure modes will depend on a number of factors, such as:

- The structure density (also known as frequency);
- Geometrical characteristics such as persistence, orientation and spacing;
- Surface properties;
- Rock mass strength; and
- Their interaction with environmental factors such as rainfall events.

Blast damage may be of concern. The particular characteristics will be both site and blast specific, and the degree to which it causes hazards will be dependent upon the success of mitigating processes (such as wall scaling). Figure 7.2 has been quoted from the work of Simmons (2003) and summarises the structures of interest.

Structures within the coal seam itself (cleats) may lead to minor instabilities in highwalls if unfavourably oriented, although typically this will not be the case. The

Figure 7.3 Unexpected highwall failures (Simmons and Simpson, 2007).

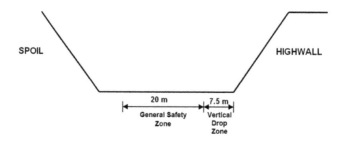

Figure 7.4 General safety and vertical drop zone of highwalls (www.qldminingsafety.org.au/_dbase_upl/1995_spk008_Petersen.pdf).

structures of interest are therefore located in overburden and interburden areas of the highwall.

Though there could be cases of unexpected highwall failures, such as in Figure 7.3, complementary methods to mitigate the risks associated with structurally controlled failures can be summarised as follows:

- Operating procedures (e.g. standoff distances from the highwall toe);
- Slope monitoring (e.g. slope stability radar); and
- Structural modelling.

If the general safety and vertical drop zones are identified at any site of operation (e.g. Figure 7.4), the risk of injury due to highwall failures can be reduced greatly by simply minimising the hazardous exposure of personnel to the minimum levels necessary for operation. It is important for all employees to know that the vertical drop zone is particularly dangerous and that the only time personnel should be allowed in this zone, with high caution, is during CHM operations and while performing essential duties such as the clean-up of the highwall toe. All other activities, e.g. general maintenance, crib, etc., should occur in the general safety zone. No one is to enter the

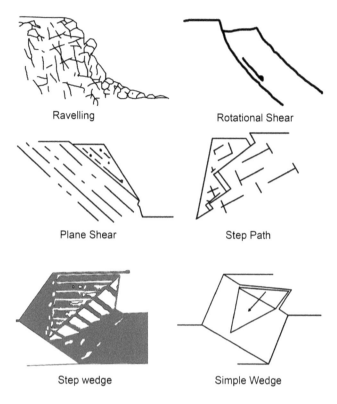

Figure 7.5 Failure modes of interest (Call and Savely, 1990).

vertical drop zone unless they are protected by 'full overhead protection' and/or have a spotter watching the highwall (QMIHSC, 1995).

7.2.1 Structural failure

The 'structural modelling' has been discussed thoroughly in Section 5.6 of Chapter 5, wherein the Discrete Fracture Network (DFN) and Polyhedral modelling, Monte Carlo sampling and the application of the Siromodel software were discussed extensively, citing case examples. To quantify the significance of discrete structures in the rock mass and predict the likelihood of structurally controlled failures, discontinuum and/or hybrid methods are required. The potential failure modes, as depicted in Figure 7.5, are categorised as:

- Ravelling;
- Plane shear;
- Simple wedge;
- Step wedge;
- Toppling; and
- Rotational shear.

It is the goal of modelling to capture as accurately as possible these potential failure mechanisms. In particular, the geometry of structures (discontinuities, fractures, faults, etc.) needs to be captured to accurately represent the potential failure volumes and locations. There are several benefits associated with structural modelling, including a better understanding of the causes of stability phenomena; better interpretation of monitoring data; and the prediction of locations and volumes of failures. Traditional analyses assume persistent structures and only simply accommodate excavation geometry. Numerical analyses are more sophisticated and account for stress propagation, fractures and potentially coupled hydro-mechanical behaviour. Polyhedral analysis is geometrically accurate but is limited to the assumption of a partitioned rock mass being an assemblage of independent blocks, usually based on input fracture network geometry. These methods are complementary and all are completely reliant on the quality of the input data defining the fracture properties.

7.3   EXAMPLES OF HIGHWALL FAILURES

In the figure below (Figure 7.6), multiple seam highwall mining appears to have caused several extensive highwall failures, endangering the working crew. In the figure, a 1.5 m thick lower seam was mined first, followed by a 0.9 m thick upper seam. A weak, laminated interburden, ranging in thickness from 1.2 to 3 m, separated the two seams. A catastrophic collapse of the web pillars occurred, which resulted in the extensive highwall failure.

Figure 7.7 shows the multiple seam highwall mining failure that resulted from the failure of web pillars where two seams were in close proximity. The highwall miner was trapped inside the entries, and was later recovered by surface excavation due to the shallow depth.

Figure 7.6 Highwall collapse in multiple seam mining area (Zipf, 1999).

Figure 7.7 Highwall collapse in multiple seam mining area (Zipf, 1999).

One of the highwall mining accidents reported by MSHA, USA in 2000 was the result of the unexpected fall of a large vertical plate of overburden strata, approximately 80 m in length and up to 3.6 m in depth, detached from a near vertical pre-existing fracture parallel to the highwall, which fatally injured one front-end loader operator working to remove coal from the stockpile at the highwall mine. The fall was believed to be initiated by the large deformation of the mine pillars under it.

7.4   IMPACT OF NEARBY BLASTING ON HIGHWALL ENTRIES AND SLOPES

Inadequate blast design can have a major impact on mining costs and may also contribute to hazardous or unstable excavated slope conditions. Slope angles and bench layouts can have a large influence on the effectiveness of blast design, so geotechnical design recommendations for slope profiles should include consideration of the capabilities and limitations of blasthole drilling rigs and site work practices (Simmons, 2013).

Any blasting activity close to highwall mining may affect the stability of highwall mining entries and web pillars, as well as the highwall of an opencast mine from where the workings are to be executed. Ground vibrations generated from day-to-day blasting operations near the highwall mining entries can impose immature roof and side collapse, thereby trapping the continuous miner. Flyrock generated from blasting near the highwall mining can also damage the equipment used for the mining, such as launch vehicles etc., and impose safety concerns for the workers (Porathur et al., 2013). Statutory authorities in India, namely the Directorate General of Mines Safety (DGMS), have stipulated the threshold values of vibrations from blasting for the safety of roofs and pillars in underground coal mines (DGMS Technical Circular No. 06, 2007).

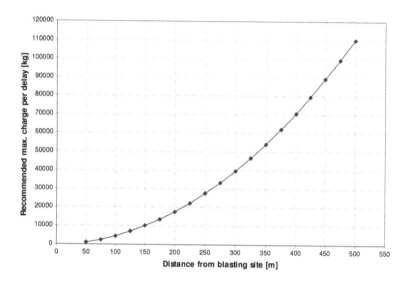

Figure 7.8 Recommended charge per delay to be fired in a round of blast for the safety of highwall mining entries (reproduction of Figure 4.66 of Section 4.4.13 of Chapter 4).

Detailed analysis on the impacts of surface blasting on highwall entries and web pillars are covered in Chapter 4, citing the different case examples of four mines. A reproduction of Figure 4.66 in Section 4.4.13 of Chapter 4 is given in Figure 7.8, which determines the charge per delay to be fired in a round of surface blast for the safety of highwall mining entries at Quarry SEB (and Quarry AB), West Bokaro of M/s Tata Steel Limited.

In Sections 4.2.24, 4.3.6 and 4.4.9 of Chapter 4, stability analyses had been carried out using the *GALENA* v 4.02 computer programme, which is based on the limit equilibrium method. The cut-off factor of safety for stability was considered to be 1.3, looking into the significance of the highwall slope. Kinematic analysis was also carried out to find the possible kinematically failured modes involving structural discontinuities.

In Medapalli Opencast Project (MOCP) of Singareni Collieries Company Limited (SCCL) stability was conducted considering circular and non-circular failure of surfaces passing through ultimate slope. The overall slope angle of the highwall was considered at 44° in both rise and dip site highwalls (Section 4.3.6 of Chapter 4). When the highwall slope is very steep, in the range of 70 to 90°, it tends to throw a higher cover pressure right at the beginning of the entry. The web pillars have three free faces at the entry mouth, which makes the webs vulnerable to spalling and crushing at the mouth of the highwall.

The results of the stability analysis obtained from the Indian case studies indicated fairly stable conditions for the highwall slopes on both the dip and rise sides with a factor of safety ranging between 1.4 and 1.75, respectively. Such values of factor of safety predominantly initiated the mine management to deploy the continuous highwall miner at the pit bottom.

7.5 BACKFILLING METHODS OF HIGHWALL MINING ENTRIES

7.5.1 Backdrop

Highwall mining methods require leaving pillars between entries to support the overburden. For conventional highwall mining, these pillars will be left in place permanently after mining finishes, which inevitably causes a loss of coal resources. One way to minimise the coal loss is to use the backfill method to assist with highwall mining operations. Backfilling the highwall mining entries is an effective way to provide support to the roofs and remnant pillars after mining. It has two major benefits:

- Significantly reduce the pillar size required for long-term stability and increase the recovery rate;
- Dispose of solid mine waste (such as overburden materials) underground and reduce the environmental impacts of its surface disposal.

A few forms of backfill may be applicable to highwall mining, including paste backfill and slurry backfill. Paste backfill requires the injection of a backfill mix made of fly ash (or crushed overburden material), cement and water in a paste form into the highwall mining entries. Once set, the backfill mix will have the required strength to support the roof and pillar. Slurry backfill requires the use of low solid concentration slurry to fill the entries, and it relies on the frictional strength of the settled solid and its stiffness to support the roof and pillar. Although the paste backfill is capable of achieving high strength and is a more effective backfill, it is a high cost method due to the requirement of cement and the filling process is much more difficult than the slurry backfill. In contrast, the slurry backfill is a more economic and operationally simpler backfill method than the other methods. If designed properly it will be able to provide sufficient support to the highwall mining entries with minimum operational costs.

To date, the backfilling methods for highwall mining are still at the stage of research and development. No routine backfill operations have yet been carried out. Since 2004, CSIRO has conducted systematic studies on new technologies for mining backfill, and has developed an economic "non-cohesive backfill" technology for both the subsidence control of operating longwall mines and the remediation of old abandoned underground coal mines (Shen and Guo, 2007; Guo et al., 2007; Shen et al., 2010a; Shen et al., 2010b; Alehossein et al., 2012; Shen et al., 2012). This technology is considered to be suitable for highwall mining backfill.

The non-cohesive backfill method is a simple method using fly ash and water as the basic backfill material. The backfill grout will be in the slurry state during the backfilling operation, which has a superb flowability. Once backfilled in the highwall mining entries, the fly ash will settle to become solid with certain frictional strength and stiffness. The settled fly ash will then provide the required support properties to the highwall entries.

The success of this backfill technology relies on several key technical aspects:

1. Grout flowability
2. Fly ash strength
3. Fly ash stiffness

4   Backfilling efficiency
5   Environmental safety of the backfill material

The following sections summarise these key technical aspects of this technology.

7.5.2 Physical and mechanical properties of fly ash

Laboratory tests were conducted to characterise and assess the physical and mechanical properties of the fly ash as a suitable and sustainable deposit for the backfilling of underground mining voids. The average particle size of the samples from several power stations in Australia is around 50 mm and the average solid particle density, or Specific Gravity (SG), is 1.8.

Deformation characteristics and the load bearing strength of settled fly ash and their variation with time, space and the surrounding environment depend on the fly ash's physical properties. An important property is density: the denser the fly ash, the stiffer and stronger it is against loading. The density of a fly ash sample depends on the mass and volume of its major components, i.e. solid particles, water and air (gas). The density of the fly ash slurry is normally in the range of $1.1–1.4\,kg/cm^3$ corresponding to a solid concentration range of 20–60%, but can reach $1.6\,kg/cm^3$ and beyond depending on the compaction and consolidation loading history and time.

7.5.2.1 Fly ash strength

The deformation and load bearing capacity of the fly ash against any potential pillar failure are the two main concerns for the suitability of fly ash as a cohesionless backfill material. Therefore, an important question arises: will the deposited fly ash have enough shear strength, characterised by friction angle and stiffness, to cater for any possible pillar volumetric expansion at any time (before or after any pillar failure). Various laboratory tests on settled fly ash were conducted at CSIRO to address both questions, including:

1   Horizontal passive strength test by a piston, where a mechanical piston of 50 mm diameter and 55 mm thickness was driven into a submerged fly ash sample inside a large cylinder. The cohesion and friction angle of the settled fly ash can be estimated from the resistance force measured.
2   Vertical bearing strength test by a footing, where a small disc footing was loaded on top of a settled and submerged fly ash. The ultimate bearing strength of the settled fly ash can be used to calculate its frictional and cohesive strength.
3   Slope test, where a slope was excavated in the submerged fly ash and the maximum stable slope angle was measured, which represents the apparent friction angle of the settled fly ash.

The results from all three types of tests suggest that the naturally settled fly ash after backfill has an internal friction angle greater than 42°. This high friction angle will be able to provide adequate support to the pillars in confined conditions.

7.5.2.2 Fly ash stiffness

The lateral deformation of a backfilled coal pillar depends on the deformation characteristics or stiffness of the backfill; the stiffer the backfill, the less lateral movement of the pillar. However, the stiffness of a saturated confined fly ash depends on the rate of loading. Loading of a saturated fly ash transfers the load immediately to the water molecules of the mix, causing the generation of excess pore pressures equal to the loading pressures and requiring time to dissipate depending on the permeability of the mix. Consolidation tests were conducted to measure the deformation behaviour and stiffness of non-cohesive fly ash slurries at different densities and loading histories – in a 50 mm diameter mould of 20 mm thickness. Although the undrained stiffness modulus of saturated fly ash is very high (around 2.2 GPa), its drained stiffness is as low as 1 MPa, depending on its density. The average stiffness expected from a normally consolidated fly ash under a 1 MPa consolidation loading is 10–20 MPa. Consolidation tests also indicate that it may take a few years for a deep (over 10 m height) column of fly ash to reach its full consolidation density in the mine.

7.5.3 Grout flowability and beach angle

The flowability of fly ash grout and its backfilling efficiency were tested by small scale laboratory backfill experiments and simulations. A small scale model was designed based on the mining layout of a room and pillar mine. The major component of the model is the rectangular testing tank with dimensions of 3 m × 2 m × 0.5 m. The tank can be tilted to model coal seam dips of 0°, 5° and 10°.

7.5.3.1 Flowability

Factors that have an influence on slurry flow behaviour include fly ash particle size, pump speed and fly ash concentration. Initial tests were carried out using fly ash under the following conditions: fly ash concentration of 45%–52%, pump speed of 100 rpm–500 rpm (corresponding to a flow rate of 0.037 l/s–0.184 l/s), tank tilt angle 10°. Under these flow conditions, the slurry (also described as grout) was successfully injected into the tank.

Other different pump speeds and fly ash concentrations were also used to test their effect on flow and the backfill process. It was found that with 65% fly ash concentration, while the injection was successful with a pump speed of 200 rpm (equivalent to a velocity of 0.092 m/s in the hose), the hose was blocked when the pump speed was reduced to 100 rpm (equivalent to a velocity of 0.046 m/s in the hose). The blockage occurred at the lowest position of the injection hole between the pump and the tank inlet. When the flow velocity of the water was reduced below a threshold, separation and deposition of fly ash particles occurred at the lower portion of the hose.

7.5.3.2 Fly ash settlement surface profile and beach angle

A major difference between a low solid concentration and non-cohesive fly ash and a high solid concentration and cohesive (with cement) fly ash slurry is the difference in their flow profile angles. The steeper the angle, the more difficult the injection into the gentle dipping highwall mining entries and the greater the risk of blockage and unfilled

(a) (b)

Figure 7.9 Settlement lines during two injection stages. (a) early stage; (b) late stage.

voids. During the injection, fly ash particles deposit gradually and continuously, normally at very low profile angles. Once the injection is disrupted or stopped, the settled, deposited particles will form a fly ash layer with a distinct settlement line or surface.

During the injection test, it was observed that the profile pattern and beach angle of the settlement line are different in different backfill stages. Figure 7.9(a) displays the profile in the early backfill stage where the deposited fly ash layer is small. The ash surface is parallel to the bottom plate of the tank at the higher part of the tilted tank. There the deposited fly ash layer is thin (less than 100 mm), and the profile beach angle is approximately equal to the tilt angle of the tank. This may be attributed to the boundary effects from the floor of the tank. At the lower part of the tank, where the fly ash layer becomes thicker (greater than 100 mm), the profile becomes nearly horizontal, whereby the beach angle is less than 1°.

As the backfill process continues and the deposited fly ash layer increases in thickness, the settlement surface profile approaches the horizontal, see Figure 7.9(b). Close to the injection hole the local deposit was influenced by the input flow forming a bump in the profile. It was observed during the 50% solid concentration backfill simulations that, regardless of the tilt angle of the tank (5 or 10°), the settlement surface becomes flatter (beach angle < 1°) once the deposited fly ash layer is thicker than 150 mm.

7.5.4 Effect of backfill

A systematic numerical study was carried out to investigate the strength of pillars before and after backfill. The effects of backfill on pillar strength using non-cohesive fly ash include:

- Increase immediate pillar strength and reduce the rate of pillar strength degradation from pillar spalling due to exposure to the environment, and roof instability that may increase the pillar working height;
- Change the post-peak pillar behaviour from rapid strength reduction to hardening so that pillars can continue to carry a significant load even after yielding;
- Reduce entry void space and hence significantly reduce the magnitude of surface subsidence from the unlikely event of panel failure.

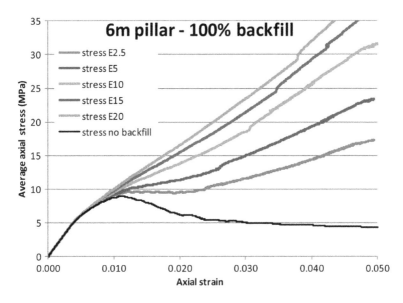

Figure 7.10 Predicted full loading curve of a 6 m pillar with or without fly ash backfill.

A single square pillar was investigated using a numerical method to demonstrate the effect of backfill (Wang et al., 2011). Although the pillar geometry is different from that of highwall mining pillars, the results are very relevant to highwall mining backfill. Based on the strength and stiffness test results, the backfill is assumed to have a Young's modulus of 20 MPa, Poisson's ratio of 0.15, zero cohesion and a friction angle of 42°.

The effect of backfill stiffness was also investigated numerically. Young's modulus of 2.5 MPa to 20 MPa was used, and the comparative results for different Young's modulus are shown in Figure 7.10. The pillar failure mode changes from a strain-softening mode (in the absence of a backfill) to a strain hardening mode (in the presence of a backfill). In other words, the yield strength of the backfill material increases with increasing deformation in a confined space; the more pillar lateral deformation, the stiffer the fly ash and the higher is the pillar yield capacity.

7.5.5   Fly ash chemistry and leaching potentials

The use of fly ash has many benefits, including its availability in large quantities from power stations near the mine sites, its ability to be made in a slurry form and its good settled strength. However, unlike the natural mine waste rocks that can readily be returned underground, the use of processed materials like fly ash requires extensive evaluation in order to prevent the contamination of groundwater or surface water supplies.

Extensive investigations were carried out on the chemistry and leaching potential of two fly ash samples collected from the Swanbank Power Station. These included (1) short-term batch tests as per the recommended standard procedure: two types of batch tests, namely, the Toxicity Characteristics Leaching Procedure (TCLP) test and the Synthetic Groundwater Leaching Protocol (SGLP) test were conducted. (2) Long-term column studies to simulate the actual mine site conditions were also carried out. Detailed tests setup can be found in Shen et al. (2012).

Overall, from the elemental concentrations analysis and the leachate analysis from the batch tests, both samples showed that the element concentration levels were generally lower than the ANZECC (Australian and New Zealand Environment Conservation Council) and QWQ (Queensland Water Quality) guidelines. However, pH of the solution had a significant effect, especially for aluminium, molybdenum and vanadium. Column studies on the two fly ash samples were carried out for a 7–19 day period and the effect of water recirculation was also evaluated. Elements such as molybdenum, selenium, vanadium, cobalt, nickel and boron were monitored closely for their presence in the leachate. Under the given test conditions, a general trend of gradual reduction in the element concentration was noted and equilibrium times for each element were different. Concentrations of organic compounds in the leachate were well below the recommended guideline levels.

7.5.6 Summary

Non-cohesive backfill is a low cost and easy-to-operate backfill technology, which can be used for highwall mining backfill. Based on the study results described in this section, it has the following key characteristics.

With a solid concentration of about 50%, fly ash grout has an excellent flowability and very low viscosity. It is capable of penetrating and filling almost any voids underground if designed properly and settling as a reasonably stiff solid to provide support to the pillars. Both the friction angle and stiffness of fly ash are a function of its density. Several different types of strength tests proved that a consolidated fly ash with a saturated density at and above 1.5–1.6 g/cc should exhibit a friction angle above 42°. Furthermore, the fly ash should produce a loading stiffness in the range of 5–15 MPa and an unloading, or over-consolidated, stiffness in the range of 10–80 MPa.

Three-dimensional numerical modelling on the interaction between fly ash backfill and underground pillars has shown that fly ash backfill to 90% roadway height is predicted to raise the pillar strength by more than 60%, or the Factor of Safety (FoS) of a marginally stable area to above 1.6, which is the number often used in rock engineering design for long-term stability.

Chemistry and leachate analysis of representative fly ash samples from a local power station showed that the elemental concentrations in the fly ash solid samples were lower than the allowed contaminant threshold and specific contaminant concentration levels. The leaching potential of the fly ashes was evaluated through a series of batch and column tests as recommended in the Australian Standards. In both types of tests, it was found that the element concentration levels were generally lower than the guidelines. These tests demonstrated that fly ash backfill is an environmentally friendly solution for the abandoned mine remediation.

Chapter 8

Design and operational guidelines

8.1 INTRODUCTION

Geo-mining conditions in highwall mining situations can often be quite complex. Multiple seams can exist with different parting thickness. Sometimes they are contiguous, having an interburden parting of less than 9 m. In such cases multi-seam interaction becomes a real challenge during mining. Seams often dip steeply and the working depth may vary from shallow to deep. In the Indian case example discussed in Chapter 4 – viz. OCP-II of Singareni Collieries Company Limited (SCCL) – the working depth of the open-pit was as high as 270 m. While working in the Medapalli Opencast Project (MOCP) of SCCL, the stone band occurrence prohibited web cut drivages to full penetration depth at many places and resulted in frequent replacement of cutting picks. Existence of hard coal, thinning of the seam, weak immediate roof layer and working in apparent dip direction were other such issues that needed special attention while designing web pillars. In such a complex environment, meticulous planning is, therefore, an utmost requirement. Hence on the basis of the detailed investigations described in the previous chapters and the performance analysis of past highwall mining case studies, guidelines have been framed for safe highwall mining operations. DGMS (India) has also issued one technical circular on the use of highwall mining technology in India based primarily on the data generated by CSIR-CIMFR and CSIRO (DGMS Technical Circular No. 06 of 2013).

8.2 WEB PILLAR STRENGTH ESTIMATION

The most crucial part of highwall mine design is to estimate the strength of the web pillar and the load on it more accurately. In this regard local expertise developed in various countries may be given due importance.

For Indian coalfields it is prudent to use the modified CSIR-CIMFR equation for web pillar strength estimation. It has been successfully used in the MOCP and the Opencast Project-II (OCP-II) of SCCL. In the modified equation (3.5) equivalent pillar width (W_e) should be considered equal to $2W$ based on the study by Wagner (1974). Extensive numerical modelling and laboratory testing revealed that for very slender pillars, with $W/h < 1.0$ long pillar strength is found to be less than the equivalent square ones, and therefore it is suggested that the estimated strength from the CSIR-CIMFR modified equation is reduced by 20% to arrive at the correct web pillar width.

Similarly the CSIRO developed web pillar strength (Equation 3.9) and the modified Mark-Bieniawski (Equation 3.4) would be most suitable for Australian and US coalfields, respectively. For other parts of the world, a suitable method from the above three equations combined with the local expertise, if any, may be put to use.

A review of catastrophic collapse and panel failures in the USA and Australia reveals that in most cases the W/h ratio of web pillar was less than 1.0. This is supported by findings of laboratory investigations where bursting phenomenon is found to be associated with a W/h of 0.5. Therefore, it is recommended that web pillars with a W/h ratio greater than or equal to 1.0 are designed for greater stability and to avoid pillars bursting over time. However, in the case of extractions at very shallow cover or with a backfilling method for web cuts, W/h ratios less than 1.0 may be designed after further scientific evaluation. If web pillars of W/h ratio less than 1.0 are used, barrier pillars of higher W/h ratios should be designed after a designed panel span using numerical modelling to avoid domino type large-scale collapses.

In addition:

- Site-specific scientific study should be undertaken in multiple seam scenarios involving empirical know-how as well as numerical modelling analysis.
- If geotechnical conditions are found to be quite different during extraction the web pillar strength should be reassessed.
- Design width of web pillars should consider support to the highwall to avoid slope failures.

Three-dimensional modelling of web pillars reveals that the long pillar is a true case of plane-strain condition, and by considering maximum depth of the panel, the strength of the web pillar can also be computed using numerical modelling by site-specific plane-strain models.

Current numerical modelling studies have revealed that for long web pillars W/h ratios greater than 3.0 are 'indestructible'. While stress-strain curves of such pillars do experience some yielding at the sides a solid core remains un-yielded and continues to take more load without eventually collapsing. Hence, from a stability requirement and mineral conservation point of view it is better to design web pillars with a W/h ratio not greater than 3.0. In most highwall mining scenarios a safety factor of 2.0 for web pillars can generally be achieved at W/h ratios less than 3.0.

8.3   ESTIMATION OF LOAD ON WEB PILLAR

After estimation of pillar strength, determination of load on pillars is the second step in pillar design. To determine load on a web pillar, the tributary area approach is still widely utilised in highwall mining. It is given by:

$$P = \frac{\gamma H (W + W_c)}{W} \tag{8.1}$$

where P is load on web pillar, MPa; γ is unit rock pressure, MPa/m; H is overburden depth, m; W_c is web cut width, m and W is width of web pillar, m.

However, in highwall mining, the load 'P' is usually not uniform throughout the length of web pillars mainly on account of varying depth caused by seam inclination, slope of highwall and other induced stresses brought about by open-pit mining.

Hence it is recommended to use Szz_{max} obtained from site-specific numerical modelling in estimation of the maximum stress over web pillars.

The hybrid empirical and numerical approach recommended for estimation of maximum stress on web pillar is therefore given by:

$$P_{max} = \frac{Szz_{max}(W + W_c)}{W} \tag{8.2}$$

where, P_{max} – maximum vertical stress on the web pillar, MPa; Szz_{max} – maximum vertical stress acting on the seam obtained from numerical model, MPa; W_c – web cut width, m and W – web pillar width, m.

8.4 DETERMINATION OF FACTOR OF SAFETY

It is usually very common for the mining community to interpret a safe mining situation in terms of Factor of Safety (FoS); this is computed from the ratio of the strength of the pillar to the load on it.

$$Safety\ factor = \frac{Strength\ of\ pillar\ (S)}{Load\ on\ Pillar\ (P)} \tag{8.3}$$

For mines in the USA Mark-Bieniawski suggests a design safety factor of 1.3–1.6 for web pillars while CSIRO in Australia suggests an FoS of 1.3 in general, and of 1.2 as critical (Duncan Fama et al., 1999) based on Local Mine Stiffness (LMS) Theory (Adhikary et al., 2002). For Indian geo-mining conditions, it is recommended to utilise a minimum FoS of 1.5 in multiple seam scenarios and a minimum FoS of 2.0 in those areas where important surface structures are to be protected from subsidence. When the web pillars are designed with a minimum of 1.5 FoS, there should not be a requirement for wider barrier pillars. If slender web pillars with W/h ratio <1.0 or with FoS less than 1.5 are used, it is necessary to leave barrier pillars after a specific number of web cuts. The width and intervals of such barrier pillars need to be determined using numerical simulation.

8.5 MULTI-SEAM MINING

Multiple extractable seams may exist at several highwall mining operation sites. For the case studies from India on the MOCP and OCP-II, multi-seam interaction was assessed beforehand using numerical modelling studies and workings were superimposed wherever the seams are in contiguity (when the interburden between two targeted coal seams is less than 9.0 m as per Indian coal mining regulations). No parting failure has been experienced so far in India, but a few panel failures have taken place in the USA due to parting collapse, especially where the multiple seam extractions were very close to each other. Empirical equations on pillar strength do not incorporate this phenomenon.

When the seams are contiguous, the interaction between the proposed extractions in both seams needs to be evaluated using numerical modelling. If the proposed web cuts in both seams are found to influence each other, it is recommended to superimpose the workings in both of the contiguous multi-seams. In such cases the web pillar width and web cut width also need to be of the same dimensions in both the seams.

In the West Bokaro mine two seams, namely seam XU and seam XL, have a thickness of 1.4 m and 1.2 m respectively, with a very thin parting of <2 m in between. At places these two seams occur as one. In this case it is difficult and dangerous to design and extract them separately. To prevent the loss of coal in any seam combined web pillar is designed, considering the combined height of extraction including the parting. Although experience is lacking in this regard it appears to be the most practical solution to recover the deposit. Hence for those seams with <3 m parting of weak strata it is recommended to create a combined design of web pillars considering the combined height of the two seams including the parting in the empirical equation. In such cases the sequence of mining should be top-down to prevent the cutter falling into the bottom seam. In such low parting multiple seam extraction, stability assessment of the combined height web pillar should also be corroborated through numerical modelling.

8.6 HIGHWALL STABILITY

Stability of a highwall is greatly influenced particularly by web pillar stability and in general by slope stability. Most of the serious accidents reported in highwall mining cases worldwide, especially those involving fatalities, were caused by slope failure triggered by multiple web pillar collapses. In India, the slope of the final highwall is kept quite flat at about 40–45° but when trench highwall mining is practised the slope angle is usually steeper in the range of 70–80°. Any mistake in the design of such steep slopes may jeopardise the safety and investment. Mostly slope stability analyses are done by quick methods such as limit equilibrium, by assuming a plane-strain scenario. However such approximations will not incorporate the highwall mining web cuts and its influence on the overall slope stability. Therefore the best practice is to pre-assess the highwall slope stability by incorporating the proposed multiple/single seam extractions through three-dimensional numerical modelling. For a suitable layout of web cuts and web pillars, damage caused by open-pit blasting to the final highwall should also be considered.

The fall of loose material from a slope face may endanger the life of miners as well as damage the CHM, and hence regular inspection of highwall should be undertaken by a certified person. Loose hazardous material from highwalls should be removed and loose unconsolidated material dressed to the angle of repose. If required wire mesh can also be provided to arrest fall.

8.7 HIGHWALL INSPECTION AND MONITORING

The first sign of instability in a highwall can be traced through the generation of fine cracks and it is therefore recommended to monitor the surface above highwalls for any kind of movement. Visual inspection of the highwall face also becomes essential and should be done regularly. Any sign of movement in the strata can help in taking suitable measures for the safety of men and machines.

Surface movements above highwalls should be monitored regularly for any sign of subsidence. Depending on the size and shape of the proposed extraction panels, subsidence survey pillars should be fixed prior to extraction. The movements (horizontal as well as vertical) of these survey points should be monitored regularly using survey instruments such as Total Station.

At a shallow depth of cover, the strata movement from the roof of the seam to the surface may also be monitored using multi-point borehole extensometers installed from the surface.

In the case of benched highwalls, the slope benches should also be monitored regularly using survey points installed over the benches, and read using a Total Station.

Highwall slope face movements should be monitored remotely during extraction using suitable survey equipment and if possible, using reflectors installed on the slope face (if the benches were accessible at some point in time). If it is economically feasible – especially if the highwall mining is planned on a very large scale – Slope Stability Radar (SSR) may also be used.

8.8 IDENTIFICATION OF GEOLOGICAL AND STRUCTURE FEATURES

Detailed geotechnical mapping of the area demarcated for highwall mining is necessary in order to locate sensitive water bodies, underground workings, hard stone bands and existing faults in the working zone. While working on the OCP-II and MOCP, full penetration was often not achieved due to the presence of unexpected hard stone bands in the middle of the coal seam. Prior knowledge on the existance of such bands might not disturb technical and economical aspects of the site. If bands are cutting across the deposit and are only few metres in thickness, although it can be cut it will definitely overload the motor of CHM. Performance and productivity of CHM is greatly affected by such geological disturbances. Hence detailed geotechnical mapping of the site is essential.

In the case of any important surface features, such as a river flowing nearby the highwall mine boundary, the end of the web cut should be designed at a sufficient distance away from the riverbank considering the angle of draw (25° in Indian coalfields) and a minimum distance of 15 m.

If old underground workings exist in a highwall mining zone, highwall drivages should be made within a safe distance from the old workings to avoid the possibility of dangers such as inundation or other hazards from the old workings. In India, a distance corresponding to one pillar width as per Reg. 99 of CMR 1957 will be considered safe enough for highwall mining operations.

Sufficient barriers should be left (minimum 15 m on either sides) against any major fault in the highwall. The overall stability incorporating the fault may also be re-evaluated using numerical modelling. In case of occurrence of unforeseen major geological changes during extraction, the design should be reviewed.

8.9   ENTRY STABILITY

Highwall mining entries are unmanned and only require temporary roof stability (a maximum stand-up time of about two days). Highwall web cuts are also of a smaller span (maximum 3.5 m), which will also ensure better stability. However, it has been found that if the immediate roof is weak clay or shale, frequent roof falls take place during extraction which causes damage to the cutter and conveyors and also results in stoppages of the machine for repair and clearing of debris. In addition, severe roof falls after mining could increase the effective height of the pillar and reduce the pillar strength, and lead to panel failure. It is therefore mandatory to assess the roof span stability using empirical methods such as *RMR* and also preferably using numerical modelling techniques.

For very weak roof scenarios, sufficient thickness of coal (say 30–50 cm) may be left in the roof while the cutter enters the web cut to prevent occurrence of frequent roof falls. This left out coal may be scrapped, if practically possible, while the cutter retreats from the web cuts, or if a second pass is done for thicker seams. If roof fall is anticipated quite consistently, the web pillar height should consider this weak layer and the web pillar designed accordingly.

8.10   SEQUENCE OF MULTIPLE SEAM EXTRACTION

In case of extraction of exposed seams from an open-pit at the final highwall, an ascending sequence of extraction should be followed, starting with the bottom most seam at the pit bottom and thereafter backfilling of the pit by forming benches using overburden dump material or any other incombustible available waste material to approach the upper seams, and so on and so forth. However, in the case of trench mining ascending as well as descending, sequence of extraction can be followed as per the ease of operation.

8.11   SEAM GAS

The mouth of web cuts should be sealed by incombustible rock (overburden dump material) within two days of completing the drivages operation in the case of a Degree I mine (please refer Table 4.2 of Chapter 4 for classification of gassiness). In cases of Degree II and III mines, arrangement should be made for inertisation of the web cut by nitrogen flushing or other such means to prevent any kind of gas explosion.

8.12   PLANS AND SECTIONS

  i.   Area demarcated for highwall mining should be marked on mine working plan.
 ii.   Accurate and detailed up-to-date hole completion plan should be maintained for each and every seam targeted for highwall mining.
iii.   Ground control plan for highwall stability should be maintained.

iv.   Hand plan on suitable scale should be prepared and provided to supervisory staff.
v.   Safety management plan should be maintained based on the risk involved.

8.13 HIGHWALL MINING EQUIPMENT

i.   All electrical enclosures and operating sensors must be intrinsically safe or flameproof.
ii.   Preventive maintenance schedule should be planned and implemented for all equipment.
iii. Every moving part of the machine should be adequately covered, fenced and guarded and no person should be allowed to repair or shift the machine in motion.
iv.   Operator should ensure proper functioning of all sensors and on-board cameras prior to the commencement of web cut.

8.14 ADDITIONAL SAFETY REQUIREMENTS

In addition to the specific requirements of a site there are a few general requirements that must be fulfilled for safe mining conditions. These are as follows:

- No person shall be allowed to enter the web cuts.
- Meticulous planning should be carried out to keep the highwall available so as to prevent the machine being idle.
- Working platform of CHM formed by dump material should be thoroughly consolidated for safe positioning and operation of the equipment.
- In no case should the bench width be less than the maximum dimension of the launch vehicle plus an additional 10 m for operational requirements, free movement of loaders, etc.
- Pit floor should be kept free from water and, if required, a suitable pumping arrangement should also be made.
- Effective drainage system should be implemented to divert surface run-off/rainwater entering into web cuts.
- If highwall slope is steep and height is more than 50 m, catch benches should be provided on the highwall at regular intervals to arrest the fall of loose material, hence ensuring stability during mining.
- Accurate and detailed up-to-date hole completion plan should be maintained for each and every seam targeted for highwall mining.
- All electrical enclosures and operating sensors should be intrinsically safe or flameproof certified.
- Preventive maintenance schedule should be planned and implemented for all equipment.
- Every moving part of the machine should be adequately covered, fenced and guarded.

- Operator should ensure proper functioning of all sensors and on-board cameras prior to the commencement of a web cut.
- No person other than authorised person should be allowed to enter into the mining area.
- Adequate arrangements should be made for the training of personnel regarding drivages of web cuts and maintaining safe operating conditions near the highwall.

Glossary

Abutment stress: Abutment means the areas of un-mined rock at the edges of an extracted area that may carry large regional loads. The peak stress at such abutments is called abutment stress.

Angle of internal friction: Also known as friction angle is a measure of the ability of a unit of rock or soil to withstand a shear stress. It is the angle, measured between the normal force and resultant force, that is attained when failure just occurs in response to a shearing stress.

Augur mining: Method for recovering coal by boring into a coal seam using augurs of circular cross section at the base of a coal strata exposed at the highwall by open pit excavation. Augering or augur mining is usually done for recovering coal from horizontal or slightly pitched coal seams for a limited depth beyond the point where open pit mining becomes uneconomical.

Barrier pillar: The pillar of coal that is left to isolate workings of one mining panel with another is known as barrier pillar.

Bedding plane: The surface that separates each successive layer of a stratified rock from its preceding layer.

CHM: Continuous Highwall Miner (CHM) is a fully automatic machine where a cutter with picks mounted on a rotating drum is taken through a conveyor into the seam almost 250–500 m deep inside to cut the coal.

Cohesion: Cohesion is the shear strength of rock when no normal stress is applied.

Compressive strength: It is the maximum stress that a material can resist under compression.

Contiguous seams: Coal seams existing at close proximity to each other (as per Indian Coal Mines Regulations, when the vertical parting is less than 9 m) are called contiguous seams.

CSIR-CIMFR: Council of Scientific and Industrial Research – Central Institute of Mining and Fuel Research.

CSIRO: Commonwealth Scientific and Industrial Research Organisation.

Density: Mass per unit volume (g/cm^3). Sandstone has a typical dry bulk density of 2.0–$2.6\,g/cm^3$.

Elasto-plastic material: A material in the state of stress between the elastic limit of a material and its breaking strength in which the material exhibits both elastic and plastic properties.

Elasto-plastic modelling: Modelling the behaviour of material that exhibits elastic deformation as well as plastic deformation under stress is termed as elasto-plastic modelling.

Empirical approach: Approach that has been developed based on experience gained by field observation or experiment.

Equivalent width: It is the term developed by many researchers to consider the shape effect in pillar design by approximating an irregular shape to a regular one.

Factor of Safety (FoS): Factor of safety is the ratio of pillar strength to stress on pillar.

Fault: A fault is a planar fracture or discontinuity in a volume of rock, across which there has been significant displacement as a result of rock mass movement.

FISH: *FISH* is a programming language embedded within *FLAC3D* software that enables user to define new variables and functions. These functions may be used to extend *FLAC3D*'s usefulness or add user defined features.

FLAC3D: *FLAC3D* is numerical modelling software and is abbreviation of Fast Lagrangian Analysis of Continua in 3 Dimensions, developed and marketed by Itasca Consulting Group Inc.

Highwall entries: Highwall entries or web cut is a rectangular cut of 50 to 500 m deep made by continuous miner in highwall to extract the coal.

Highwall mining: It is the modified version of earlier auger mining. In this method, a series of parallel drivages is made in the final highwall following seam horizon to excavate the deposit.

Highwall stability: Stability or safety factor of the final highwall slope is known as highwall stability; see "Slope stability factor".

Hybrid modelling: It is the combined approach of numerical modelling and empirical approaches to analyse the stability scenario.

In situ stress: It is the virgin stress existing in the rock mass prior to any kind of excavation.

Joint: It is a fracture dividing rock into two sections that have not moved away from each other.

Long-term stability: It refers to stable condition of excavation and its surroundings for several years even after the mining operations are over in that region.

Mineable reserve: An estimate of the economically extractable or mineable part of a mineral resource is known as mineable reserve.

Mohr-Coulomb material model: It represents material behaviour as linear yield envelope in Mohr's stress space. It divides the material in two domains. The domain below the envelope represents the material that remains un-failed and is in elastic region. The domain above the envelope represents the material where damage has occurred.

MSHA: Abbreviation of Miner Safety and Health Administration, a U.S. safety governing body.

Multi-seam extractions: When extraction is carried out in multiple coal seams that are occurring at intervals in strata it is termed as multi-seam extraction.

Numerical modelling: It is a mathematical tool by which mining conditions can be simulated with the help of rock properties, material behaviour and *in situ* stress condition.

Peak Particle Velocity (PPV): It is used for assessment of structural damage potential and refers to the maximum speed of a particular particle as it oscillates about a point of equilibrium that is moved by a passing wave resulting from blasting. The unit is mm/s.

Pillar load: Load offered by overlying strata over a pillar is termed as pillar load.

Pillar strength: It is the maximum resistance of a pillar to axial compression.

Poisson's ratio: It is the ratio of lateral strain to longitudinal strain as a result of applied forces.

RMR: RMR is the abbreviation of Rock Mass Rating. It is a rating system through which the rock mass is characterised.

Roadways: Any part of a passage or gallery below ground which is maintained in connection with mining operations.

Seam dip: It is the inclination of a coal seam from the horizontal; seam dip is always measured downwards at right angles to the strike.

Seam gas: Seam gas or Coal Seam Gas (CSG) is simply natural gas, which consists mainly of methane – a colourless and odourless gas, found in coal deposits.

Shear failure: Failure in which movement caused by shearing stresses in a soil or rock mass is of sufficiently higher magnitude in comparison to the strength to cause failure or permanent deformation of the structure.

Short-term stability: It refers to stable condition of excavation till the mining operations are over in that particular panel or zone.

Simulation: It represents the imitation of the operation of a real-world process or system over time.

Slenderness ratio: It is the ratio of width to height of a pillar (W/h ratio).

Slope stability factor: It is defined as the ratio of retaining forces to dislodging forces along a pre-defined path from toe to crust of a rock or soil slope.

Strain: It is defined as change in length to original length.

Strain-softening model: Strain softening model is useful to simulate non-linear material hardening or softening. Strain softening material resists significant strain under initial increase in stress but later due to incorporation of many fractures it undergoes continuous strain with increasing stress. After onset of yield, the material cohesion, friction and dilation are allowed to change as a function of plastic strain.

Stress: It is defined as load (or force) acting at a point per unit area. The unit is MPa.

Surface subsidence: It is the depression of ground as a result of underground mining operations.

Trench mining: When a trench is cut in the property to expose the coal seam and a series of parallel extraction drivages is made in the rise as well as dip side of the trench, the operations are termed as "Trench Mining".

Vertical stress: The stress due to overlying strata is termed as vertical stress.

Web pillar: It is the long and narrow pillar that is left adjacent to web cuts in highwall mining operations for overall stability of highwall.

W/h ratio: It is the ratio of width to the height of a pillar (slenderness ratio).

Young's modulus: It is the ratio of change in stress to corresponding strain as a result of change in stress when specimen is under uniaxial compression. The unit is GPa.

References

Adhikary, D. and Duncan Fama, M. E. (2001). Highwall mining panel stability. *CSIRO Exploration and Mining Report No. 754F.*

Adhikary, D.P., Shen, B. and Duncan Fama, M.E. (2002). A study of highwall mining panel stability. *International Journal of Rock Mechanics and Mining Sciences*, 39(5): 643–659.

Afrouz, A. A. (1994). Placement of backfill. *Mining Engineer*, February, 205–211.

AISRF Final Report: Agreement ST050173. (2015). (Baotang Shen, Shivakumar Karekal, Marc Elmouttie, John L. Porathur and Pijush Pal Roy): Highwall mining design and development of norms for Indian conditions. *CSIRO Energy and CSIR-Central Institute of Mining & Fuel Research, EP 156317/GAP/MT/DST/AJP/91/2011-12*, August.

Alehossein, H., Shen, B. and Qin, Z. (2012). Viscous, cohesive, non-Newtonian, depositing, slurry pipe flow. *International Journal of Fluid Mechanics Research*, 39(3): 191–215.

Baecher, G. B. and Lanney, N. A. (1978). Trace length biases in joint surveys. *Proc. 19th US Symposium on Rock Mechanics (USRMS), 1–3 May, Reno, Nevada, USA*, 1: 56–65.

Baecher, G. B., Lanney, N. A. and Einstein, H. H. (1977). Statistical description of rock properties and sampling. *Proc. 18th US Symposium on Rock Mechanics, Colorado School of Mines, Golden CO*, 5C1: 1–8.

Barton, N., Lien, R. and Lunde, J. (1974). Engineering classification of rock masses for the design of tunnel support. *Rock Mechanics*, 6(4): 189–236.

Bauer, A. and Calder, P. N. (1971). The influence and evaluation of blasting on stability. *Proc. 1st International Conference on Stability in Open Pit Mining, Vancouver, Canada, 23–25 November*, 83–94.

Bieniawski, Z. T. (1976). Rock mass classification in rock engineering. *Proc. of the Symposium in Exploration for Rock Engineering (ed. Z.T. Bieniawski), Vol. 1, Cape Town*, 97–106.

Bieniawski, Z. T. (1984). Rock mechanics design in mining and tunnelling. *Rotterdam: A. A. Balkema.*

Bieniawski, Z. T. (1987). Strata control in mineral engineering. *Rotterdam: Balkema*, 59–90.

Bieniawski, Z. T. (1989). Engineering rock mass classifications. *John Wiley & Sons, New York.*

Bieniawski, Z. T. (1993). Classification of rock masses for engineering: The RMR system and future trends. *In Hudson (ed.) Comprehensive Rock Engineering, Vol. 3, Rock Testing and Site Characterisation, OXFORD, Pergamon Press*, 553–573.

Billaux, D., Chiles, J. P., Hestir, K. and Long, J. C. S. (1989). Three-dimensional statistical modelling of a fractured rock mass: An example from the Fanay-Augeres Mine. *International Journal of Rock Mechanics and Mining Sciences and Geomechanics Abstracts*, 26(3/4): 281–299.

Calder, P. N. and Jackson, R. J. (1981). Revised perimeter blasting chapter. *CANMET Pit Slopes Manual, Mining Research Laboratories, Ottawa, Canada.*

Call, R. D. and Savely, J. P. (1990). Open pit rock mechanics. In B. A. Kennedy (Ed.), *Surface Mining. (2nd ed.). Proc. Society for Mining, Metallurgy and Exploration, Inc.*, 860–882.

Chiapetta, R. F. (2001). The importance of presplit and field controls to maintain stable high-walls, eliminate coal damage and overbreak. *Proc. 10th Hightech Seminar on State of the Art Blasting Technology, Instrumentation and Explosive Application*, GI-48, Nashville, Tennessee, USA, July 22–26.

Christensen, P. (2004). Highwall mining revival around the corner. Digging Deeper *AMC Mining Consultants Newsletter*, May, 1–2.

Cosserat, E. and Cosserat., F. (1909). Theorie des corps deformables. *Paris: Hermann*.

CSIR-CIMFR Report of Investigations. (2000). Slope stability at Quarries A, B and E, West Bokaro Collieries, TISCO. *GC/MT/51/2000-2001*, November.

CSIR-CIMFR Report of Investigations. (2004). Parameters for controlling ground vibration and flyrock within tolerable limits at South-Eastern Block of TISCO, West Bokaro Group of Collieries. *GC/MT/7/2004-2005*, May.

CSIR-CIMFR Report of Investigations. (2005). Advice on slope stability of ultimate mine slope and dumps of OCP-II, Ramagundam, SCCL. *GC/MT/85/2004-2005*, January.

CSIR-CIMFR Report of Investigations. (2008). Highwall mining at Ramagundam Opencast Project-II (OCP-II) of M/s Singareni Collieries Company Limited (SCCL): Phase-I study. *GC/MT/139/2007-08*, February.

CSIR-CIMFR Report of Investigations. (2008a). Scientific study to assess the impact of opencast blast vibrations at OCP-II on the stability of proposed punch entries for Adriyala Shaft Project, RG-III, SCCL. *GC/MT/113/2007-08*, September.

CSIR-CIMFR Report of Investigations. (2008b). Advice on slope stability of final highwall slope of SE Block, West Bokaro Collieries, Tata Steel Limited. *GC/MT/39/2008-09*, November.

CSIR-CIMFR Report of Investigations. (2009). Feasibility study for highwall mining at Quarry SEB (and Quarry AB), West Bokaro of M/s Tata Steel Limited. *GC/MT/174/2008-09*, March.

CSIR-CIMFR Report of Investigations. (2011). Design and feasibility assessment of highwall mining at Medapalli Opencast Project of M/s Singareni Collieries Company Limited (SCCL). *CNP/2586/2010–2011*, March.

CSIR-CIMFR Report of Investigations. (2014). *In situ* trench slope stability study at Sharda opencast highwall mining project, SECL. *CNP/N/3715/2013-14*, August.

CSIR-CIMFR Report of Investigations. (2015). Scientific study for improvement of blasting efficiency and presplit blasting for formation of highwall benches at Sharda Highwall Mining Project. *CNP/4012/2014–15*, June.

CSIR-CIMFR S&T Report of Investigations. (1987). Geomechanical classification of coal measure roof rocks vis-à-vis roof support. *CSIR-CIMFR, Dhanbad*.

Dershowitz, W. S. (1984). Rock joint systems. Ph.D. Thesis, *Massachusetts Institute of Technology, Cambridge, MA*.

Dershowitz, W., Lee, G. and Josephson, N. (2007). FracMan interactive discrete feature data analysis, geometric modelling and exploration simulation. *User Documentation, Version 7. Seattle: Golder Associates Inc., USA*.

DGMS (Tech.) (S&T) Circular No. 06. (2007). Damage of below ground structures due to blast induced vibration in nearby opencast mines. *Directorate General of Mines Safety, Ministry of Labour*, GoI, May, 1–4.

DGMS (Tech.) (S&T) Circular No. 06. (2013). Use of highwall mining technology in Indian coal mines. *Directorate General of Mines Safety, Ministry of Labour*, GoI, August.

Duncan Fama, M. E., Trueman, R. and Craig, M. S. (1995). Two- and three-dimensional elasto-plastic analysis for coal pillar design and its application to highwall mining. *International Journal of Rock Mechanics and Mining Sciences & Geomechanics Abstracts*, 32: 215–225.

Duncan Fama, M. E., Follington, I. L., Leisemann, B. E., Shen, B. and Medhurst, T. P. (1997). Geomechanics design guidelines for highwall mining under Australian conditions. Program

4 – Project Consolidation and Technology Transfer. *CSIRO Division of Exploration and Mining, Confidential Report No. 331C.*

Duncan Fama, M. E., Shen, B., Craig, M. S., Kelly, M., Follington, I. L. and Leisemann, B. E. (1999). Layout design and case study for highwall mining of coal. *Proc. 9th International Congress on Rock Mechanics (eds. G. Vouille and P. Berest)*, 1: 265–268.

Duncan Fama, M. E., Shen, B., Boland, J., Harbers, C., Carvolth, D., Leisemann, B. E. and Maconochie, A. P. (1999a). Investigation and design for Pit 20D upper for highwall mining operations at Moura Mine, QLD. *CSIRO Exploration and Mining Report No. 605C.*

Duncan Fama, M. E., Shen, B., Craig, M. S., Kelly, M., Follington, I. L. and Leisemann, B. E. (1999b). Layout design and case study for highwall mining of coal. *Proc. 9th International Congress on Rock Mechanics (eds. G. Vouille and P. Berest), Balkema, Rotterdam*, 1: 265–268.

Duncan Fama, M. E., Craig, M. S. and Coulthard, M. A. (1999c). Factors affecting design of highwall mining pillars: Parametric studies. *CSIRO Exploration and Mining Report No. 617F.*

Duncan Fama, M. E., Shen, B. and Maconochie, P. (2001). Optimal design and monitoring of layout stability for highwall mining. *CSIRO Exploration and Mining Report No. 887F.*

Elmouttie, M., Poropat, G. and Hamman, E. (2009). Simulations of the sensitivity of rock structure models to field mapped parameters. *In: Slope Stability 2009, Santiago, Chile*, 1–9 November.

Elmouttie, M., Poropat, G. and Krähenbühl, G. (2010). Polyhedral modelling of rock mass structure. *International Journal of Rock Mechanics and Mining Sciences*, 47(4): 544–552.

Elmouttie, M. K. and Poropat, G. V. (2012). A method to estimate *in situ* block size distribution. *Rock Mechanics and Rock Engineering*, 45(3): 401–407.

Elmouttie, M., Poropat, G. and Soole, P. (2012). Open-cut mine wall stability analysis utilising discrete fracture networks. *ACARP Project C20029 Final Report. CSIRO Earth Science and Resource Engineering EP122391.*

Elmouttie, M., Krähenbühl, G. and Poropat, G. (2013). Robust algorithms for polyhedral modelling of fractured rock mass structure. *Computers and Geotechnics*, 53: 83–94.

Esterhuizen, G. S. (2006). An evaluation of the strength of slender pillars. *SME, preprint, 06-003.*

Follington, I. L., Duncan Fama, M. E., Craig, M. S., Leisemann, B. E. and Medhurst, T. P. (1994). Highwall mining project report on factors affecting highwall mining design at Pit 17 DU South Moura Mine, QLD. *CSIRO Exploration and Mining Report No. 103C.*

Furukawa, H., Matsui, K., Sasaoka, T. and Shimada, H. (2009). Application of a punch mining system to Indonesian coal mines, *Proc. 2nd International Symposium of Novel Carbon Resources Science, Earth Resource Science and Technology, Bandung, Indonesia*, III1–III9.

Galvin, J. M., Hebblewhite, B. K. and Salamon, M. D. G. (1999). University of New South Wales coal pillar strength determinations for Australian and South African mining conditions. *2nd International Workshop on Coal Pillar Mechanics and Design, NIOSH IC9448*, 63–71.

Giacomini, A., Thoeni, K., Lambert, C., Booth, S. and Sloan, S. W. (2012). Experimental study on rockfall drapery systems for open-pit highwalls. *International Journal of Rock Mechanics and Mining Sciences*, 56: 171–181.

Gosset, W. S. (1908). Student-t Test (Cited from Zabell, S. L. (2008). On Student's 1908 article – "The Probable Error of a Mean". *Journal of the American Statistical Association*, March, 103(481): 1–7).

Grimstad, E. and Barton, N. (1993). Updating the Q-System for NMT. *Proc. International Symposium on Sprayed Concrete – modern use of wet mix sprayed concrete for underground support, Fagernes, Oslo: Norwegian Concrete Association*, 46–66.

Guo, H., Karekal, S., Poropat, G. and Soole, P. (2012). Pit wall strength estimation with 3D imaging. *ACARP-C18030, CSIRO Report No. EP132764.*

Guo, H., Shen, B. and Chen, S. (2007). Investigation of overburden movement and a grout injection trial for mine subsidence control. *Proc. 1st Canada-U.S. Symposium on Rock Mechanics: Meeting Society's Challenges and Demands (eds. Eberhardt, Stead & Morrison), May, Vancouver*, 2: 1559–1566.

Gustafsson, R. (1973). Swedish Blasting Techniques, *SPI Publishing, Gothenburg, Sweden*.

Hagan, T. N. and Mercer, J. K. (1983). Safe and efficient blasting in open-pit mining. *Proc. Workshop held by ICI Australian Operation Limited, Karratha*, 23–25 November.

Hawley, J. (www.prweb.com/releases/2008/06/prweb992884.htm) Push/pull auger conveying system (https://mining.cat.com/cda/files/2884582/11/, accessed on 1/5/15).

Highwall Mining Services (HMS). (1997). CHM panel design, Pit B, Yarrabee Mine. *Report prepared for Yarrabee Coal Company Limited*.

Hoek, E. (2012). Blast damage factor D. Technical Note for RocNews. February 2, *Winter 2012 issue, RocScience*, 1–7. (Available from: http://www.rocscience.com/assets/files/uploads/8584.pdf).

Hoek, E. and Brown, E. T. (1980). Empirical strength criterion for rock masses. *Journal of Geotechnical Engineering Division, ASCE*. 106(GT9): 1013–1035.

Hoek, E. and Bray, J.W. (1981). Rock slope engineering, *3rd edition, Institute of Mining and Metallurgy, London*.

Hoelle, J. (2003). Analysis of unsupported roof spans highwall mining at Moura coal mine. *Proc. Coal Operators' Conference, (ed. N. Aziz), University of Wollongong and the Australasian Institute of Mining and Metallurgy, Australia*, 1–5.

Huddlestone-Holmes, C. R. (2005). The essence of FIA: A study of the distribution of foliation intersection axes data and its significance from hand sample to regional scales. *Ph.D. Thesis, James Cook University, Townsville, Australia*.

Huddlestone-Holmes, C., Elmouttie, M., Adhikary, D. and Balusu, B. (2008). Punch entry highwall mapping and stability study using Sirovision™ and Siromodel produced for Singareni Collieries Company Limited. *CSIRO Project Report*.

Huddlestone-Holmes, C. R. and Ketcham, R. A. (2010). An X-ray computed tomography study of inclusion trail orientation in multiple porphyroblasts from a single sample. *Tectonophysics*, (doi: 10.1016/j.tecto.2009.10.021), 480: 305–320.

Hustrulid, W. A. and Swanson, S. R. (1977). Coal pillar strength study. *USBM OFR*, 166–177.

Iannacchione, A. T. (1999). Analysis of pillar design practices and techniques for U.S. limestone mines. *Trans. Inst. Min. Metall. (Section A: Mining Industry)*, 108: 152–160.

International Society for Rock Mechanics. (1978). Commission on standardization of laboratory and field tests. Suggested methods for the quantitative description of discontinuities in rock masses. *International Journal of Rock Mechanics and Mining Sciences and Geomechanics Abstracts*, 15: 319–368.

ISEE Blasters' Handbook: 18th Edition. (2011). *International Society of Explosives Engineers, USA*. 878–884.

ISRM. (1978). Suggested methods for the quantitative description of discontinuities in rock masses. *Commission on the Standardization of Laboratory and Field Tests in Rock Mechanics*, 319–368.

Itasca. (1994). 3DEC – Three-dimensional distinct element code, version 1.5. *User's Manual, Itasca Consulting Group, Inc., Minneapolis, Minnesota 55401, USA*.

Itasca. (1995). FLAC – Fast Lagrangian analysis of continua, version 3.3. *User's Manual, Itasca Consulting Group, Inc., Minneapolis, Minnesota 55401, USA*.

Itasca. (1996). FLAC3D – Fast Lagrangian analysis of continua in three-dimension, version 1.1. *User's Manual, Itasca Consulting Group, Inc., Minneapolis, Minnesota 55401, USA*.

Itasca. (1996). UDEC – Universal distinct element code, version 3.0. *User's Manual, Itasca Consulting Group, Inc., Minneapolis, Minnesota 55401, USA*.

Itasca. (1999). *PFC2D* – Particle flow code in 2 dimensions, version 2.0. *User's Manual, Itasca Consulting Group, Inc., Minneapolis, Minnesota 55401, USA.*

Jawed, M., Sinha, R. K. and Sengupta, S. (2013). Chronological development in coal pillar design for bord and pillar workings: A critical appraisal. *Journal of Geology and Mining Research*, January, 5(1): 1–11.

Jethwa, J. L. (1981). Evaluation of rock pressures in tunnels through squeezing ground in lower Himalayas. *Ph.D. Thesis, Department of Civil Engineering, University of Roorkee, India.*

Jha, N. C. (2010). Policy support for mitigation of constraints for increased production. *Proc. 3rd Coal Summit on Coal for Economic Growth, New Delhi, India*, 24–25 September.

Jimeno, L. C., Jimeno, L. E. and Carcedo, A. J. F. (1995). Drilling and blasting of rocks. *Rotterdam, The Netherlands: Balkema.*

Jing, L. (2000). Block system construction for three-dimensional discrete element models of fractured rocks. *International Journal of Rock Mechanics and Mining Science and Geomechanics Abstracts*, 37: 645–659.

Jing, L. and Stephansson, O. (1994). Topological identification of block assemblages for jointed rock masses. *International Journal of Rock Mechanics and Mining Sciences and Geomechanics Abstracts*, 31(2): 163–172.

Karekal, S., Adhikary, D. P. and Rao B. (2008). Adriyala and KTK longwall projects: Hydro-geomechanical framework and PMP characterisation. *CSIRO Report No. P2008/2412.*

Kelly, M., Duncan Fama, M. E., Shen, B., Follington, I. L. and Leisemann, B. E. (1998). Investigation of highwall mining instability, Pit 16BL South, Moura Mine, QLD. *CSIRO Exploration and Mining Report No. 467C.*

Kelly, M., Hainsworth, D., McPhee, R. J., Shen, B., Wesner, C., Wright, B., Balusu, R., Adhikary, D., Guo, H., Wendt, M. and Xue, S. (2001). ACARP Project C9017 – Rapid roadway development project progress report. *CSIRO Exploration and Mining Report No. 828F.*

Konya, C. J. and Walter, E. J. (1990). Surface blast design. *Prentice Hall. Inc., New Jersey, USA.*

Lin, D., Fairhurst, C. and Starfield, A. M. (1987). Geometrical identification of three dimensional rock block systems using topological techniques. *International Journal of Rock Mechanics and Mining Science and Geomechanics Abstracts*, 24(6): 331–338.

Loui, John P., Pal Roy, P., Kushwaha, A. and Rao, K. U. M. (2008). Web and barrier pillar design for highwall mining at OCP-II of SCCL, India. *Proc. of Korean Rock Mechanics Symposium on International Cooperation and Rock Engineering for Development of Mineral Resources and Infrastructures, Korea*, 21–23 October, 123–134.

Loui, John P., Pal Roy, P. and Sinha, A. (2011). Design of highwall mining at West Bokaro division of Tata Steel Limited, *Journal of the Institution of Engineers IE(I)-MN*, August, 92: 1–7.

Lu, J. (2002). Systematic identification of polyhedral rock blocks with arbitrary joints and faults. *Computers and Geotechnics*, 29: 49–72.

Mark, C. (1999). Application of coal mine roof rating (CMRR) to extended cuts. *Mining Engineering*, 51(4): 2–56.

Mark, C. and Iannachione, A. (1992). Coal pillar mechanics: Theoretical models and field measurements compared. *Proc. Workshop on Coal Pillar Mechanics and Design, US Department of Interior*, IC 9315, 78–93.

Mark, C., Listak, J. M. and Bieniawski, Z.T. (1988). Yielding coal pillars – field measurements and analysis of design methods. *Proc. 29th U.S. Symposium on Rock Mechanics, Minneapolis, MN, University of Minnesota*, 261–270.

Marsal, D. (1987). Statistics for geoscientists. *Pergamon Press, UK.*

Matsui, K., Shimada, H., Kramadibrata, S. and Rai, M. S. (2001). Some considerations of highwall mining system in coal mines. *Proc. 17th International Mining Congress and Exhibition of Turkey – IMCET2001*, ISBN: 975-395-417-4, 269–276.

Matsui, K., Shimada, H., Sasaoka, T., Ichinose, M. and Kubota, S. (2000). Highwall mining system with backfilling, *Proc. 9th Symposium on Mine Planning and Equipment Selection, Athens, Greece*, 333–338.

Mauldon, M. (1998). Estimating mean fracture trace length and density from observations in convex windows. *Rock Mechanics and Rock Engineering*, 31(4): 201–216.

McNally, G. H. (1996). Estimation of geomechanical properties of coal measure rocks for numerical modelling. *Proc. Symposium on Geology in Longwall Mining, Sydney, Coalfield Geology Council of New South Wales, Australia*, 63–72.

Medhurst, T. P. (1996). Estimation of the in situ strength and deformability of coal for engineering design *(Ph.D. thesis). University of Queensland, Australia.*

Medhurst, T. P. (1999). Highwall mining: practical estimates of coal-seam strength and the design of slender pillars, *Trans. Inst. Min. Metall. (Section A: Mining Industry)*, 108: 161–171.

Menéndez-Díaz, A., González-Palacio, C., Álvarez-Vigil, A. E., González-Nicieza, C. and Ramírez-Oyanguren, P. (2009). Analysis of tetrahedral and pentahedral key blocks in underground excavations. *Computers and Geotechnics, Elsevier*, 36(6): 1009–1023.

Molinda, G. M. and Mark, C. (1994). Coal mining roof rating (CMRR): A practical rock mass classification for coal mines. *United States Bureau of Mines Report, Pittsburgh, PA, USA, IC9387.*

MSHA. (2000). Fatal fall of highwall May 24, (2000). Report of Investigation ID No. 15-11005, MSHA District 6 Office, http://www.msha.gov/FATALS/2000/ FTL00C14.htm. accessed on 25/06/2015.

Newman, D. and Zipf, R. K. (2005). Analysis of highwall mining stability – the effect of multiple seams and prior auger mining on design. *Proc. 24th International Conference on Ground Control in Mining, Morgantown, West Virginia, USA*, 2–4 August, 208–217.

NIRM Report of Investigation. (1997). Geotechnical studies and design of slopes at Medapalli Opencast Project, RG-III Area, SCCL. *SS/95/01.*

NIRM Test Report. (2007). Test report of coal and sandstone samples of Medapalli OCP, SCCL. *NIRM/RFM/TR2/2007-2008.*

NIRM. (2008). Investigation into *in situ* strength of coal at GDK-10A Incline and GDK-10 Incline, RG-III Area, SCCL. *NIRM Project No. GC-08-01-c*, April.

Obert, L. and Duvall, W. I. (1967). Rock mechanics and the design of structures in rock. *Wiley, New York.*

Olofsson, S. O. (1998). Applied explosives technology for construction and mining. *Rotterdam, The Netherlands: Balkema.*

Pal Roy, Pijush. (2005). Rock blasting effects and operations. *New Delhi: Oxford and IBH Publishing Company Pvt. Ltd. (also published by A. A. Balkema, Rotterdam, The Netherlands).*

Palmstrom, A. (2006). Use and misuse of rock mass classification systems with particular reference to the Q-system. *Tunnels and Underground Space Technology*, 21: 575–593.

Park, H. J. and West, T. R. (2002). Sampling bias of discontinuity orientation caused by linear sampling technique. *Engineering Geology*, October, 66(1): 99–110.

Persson, P. A., Holmberg, R. and Lee, J. (1994). Rock blasting and explosive engineering. *Boca Raton, Florida, USA: CRC Press.*

Porathur, J. L., Karekal, S. and Pal Roy, P. (2013). Web pillar design approach for highwall mining extraction. *International Journal of Rock Mechanics and Mining Sciences*, 64: 73–83.

Poropat, G., V., Mamic, G., J., LeBlanc Smith, G. and Soole, P. (2004). Highwall hazard mapping project. *ACARP Project C9036, CSIRO Exploration and Mining Report No. 1203F (Open).*

QMIHSC. 1995. *Queensland Mining Industry Health and Safety Conference.* Available at: http://www.qldminingsafety.org.au/_dbase_upl/1995_spk008_Petersen.pdf. Accessed on March 2017. 1–7.

Read, J. and Stacey, P. (2009). Guidelines for open pit slope design. *CRC Press (ISBN 9780415874410)*, November.

Roberts, D. P., Ryder, J. A. and van der Merwe, J. N. (2005). Development of design procedures for long slender pillars. *Final Report Task 2.14, Coaltech 2020*, 11 March.

Rogers, S. F., Kennard, D. K., Dershowitz, W. S. and Van, As. A. (2007). Characterising the in-situ fragmentation of a fractured rock mass using a discrete fracture network approach. *In Eberhardt, E., Stead, D. and Morrison, T. (eds), Rock mechanics: Meeting society's challenges and demands. London: Taylor & Francis Group.*

Ryder, J. A. and Özbay, M.U. (1990). A methodology for designing pillar layouts for shallow mining. *ISRM Symposium on Static and Dynamic Considerations in Rock Engineering, Swaziland.*

Sakaguchi, H., Mühlhaus, H. B. and Wei, Y. (1999). Modelling dynamic fracture and flow in jointed rock. *Explo 99, Kalgoorlie*, 7–11 November, 119–125.

Salamon, M. D. G. and Munro, A. H. (1967). A study of the strength of coal pillars. *Journal of the South African Institute of Mining and Metallurgy*, 68(2): 55–67.

Salamon, M. D. G., Galvin, J. M. and Hebblewhite, B. K. (1998). Development of an integrated pillar strength determination based on Australian and South African case histories. *Proc. Int. Conf. Geomechanics/Ground Control in Mining and Underground Construction, The University of Wollongong*, 58–67.

Sartaine, Jeff (1993). Highwall mining in Australia. *President, Mining Technologies Inc., Australia (Notes).* [www.qldminingsafety.org.au/_dbase_upl/1993_spk016_Sartaine.pdf]. accessed on 10/05/2015.

Sasaoka, T., Hamanaka, A., Shim, Hideki., Matsui, K., Takamoto, H., Meechumna, P. and Laowattanabandit, P. (2013). Study on application of highwall mining system under weak geological condition in Thailand, *Journal of Physical Science and Application*, 3(6): 359–365.

Sasaoka, T., Shimada, H., Hamanaka, A., Sulistianto, B., Ichinose, M. and Matsui, K. (2015). Geotechnical issues on application of highwall mining system in Indonesia. *Vietrock2015 – an ISRM specialized conference, Hanoi, Vietnam*, 12–13 March.

Seedsman, R. (1996). A review of the hydrogeological aspects of Australian longwalls. In G. H. McNally and C. R. Ward (eds.), *Proc. Symposium on Geology in Longwall Mining*, November, 17–25.

Seegmiller, B. L. (2000). Coal mine highwall stability. In W. A. Hustrulid, M. K. McCarter and D. J. A. Van Zyl (eds.), *Slope stability in surface mining. Colorado, USA: SME*, 257–264.

Seib, W. T. (1992). Advances in highwall mining. *3rd Large Open Pit Mining Conference, Mackay, Queensland*, 145–150.

Seib, W. T. (1993). Australasian coal mining practice. In Hargraves, A. J. & Martin, C. H. (eds.). *The Australasian Institute of Mining Metallurgy*, 238–242.

Shanmukha, M. R., Gudlavalleti, U. B. and Karekal, S. (2015). Estimation of uniaxial compressive strength of coal measures of Pranhita-Godavari Valley, India using sonic logs. *Proc. 15th Coal Operators' Conference, University of Wollongong, The Australasian Institute of Mining and Metallurgy and Mine Managers Association of Australia*, 36–47.

Shen, B. and Duncan Fama, M. E. (1996). Span stability prediction for highwall mining – analytical and numerical studies. *CSIRO Exploration and Mining Report No. 316F.*

Shen, B. and Duncan Fama, M. E. (1997). A laminated span failure model for highwall mining span stability assessment. *Proc. 9th International Conference on Computer Methods and Advances in Geomechanics, Wuhan, China*, November, 1587–1591.

Shen, B. and Duncan Fama, M. E. (1999). A coal roof failure model for highwall mining span stability assessment. *Proc. 8th Australia New Zealand Conference on Geomechanics, Hobart*, 2: 633–639.

Shen, B. and Duncan Fama, M. E. (1999a). A coal roof failure model for highwall mining span stability assessment. *CSIRO Exploration and Mining Report No. 610F.*

Shen, B. and Duncan Fama, M. E. (1999b). Review of highwall mining experience in Australia and case studies. *Exploration and Mining Report No. 616F.*

Shen, B., Duncan Fama, M. E., Adhikary, D. P., Guo, H., Poulsen, B., Ross, X., Luo, J., Soole, P., Roberts, G., Poropat, G. and Maconochie, A. P. (2000). Monitoring and analysis of rock mass response to highwall mining in Pit 20D Upper North, Moura mine, *CSIRO Exploration and Mining Report No. 732F, Melbourne, Australia: CSIRO Publishing.*

Shen, B. and Duncan Fama, M. E. (2001). Geomechanics and highwall mining. *World Coal,* February, 10(2): 35–38.

Shen, B. and Guo, H. (2007). A laboratory study of grout flow and settlement in open fractures. *Proc. 1st Canada-U.S. Symposium on Rock Mechanics: Meeting Society's Challenges and Demands, Vancouver,* May, 1: 747–754.

Shen, B., Alehossein, H., Poulsen, B., Huddlestone-Holmes, C., Zhou, B., Luo, X., Wang, H., Qin, J., Duan, J., Zency, Z., van de Werken, M. and Williams, D. (2010a). Collingwood park mine remediation – subsidence control using fly ash back-filling. *CSIRO Earth Science and Resource Engineering Report EP105068,* 1–227.

Shen, B., Alehossein, H., Poulsen, B and Waddington, A. (2010b). ACARP Project C16023 – Subsidence control using coal washery waste. *CSIRO Exploration and Mining Report P2010/65.*

Shen, B., Alehossein, H., Poulsen, B., Huddlestone-Holmes, C., Luo, L., Thiruvenkatachari, R., Qin, J., Duan, Y. and Williams, D. (2012). Collingwood park mine remediation – Phase 2: Technology Validation. *CSIRO Earth Science and Resource Engineering Report EP122132.*

Sheorey, P. R. (1992). Pillar strength considering *in situ* stress, *Workshop on Coal Pillar Mechanics and Design, Santa Fe., US Bureau of Mines, IC 9315,* 122–127.

Sheorey, P. R. (1997). Empirical rock failure criteria. *Rotterdam, The Netherlands: A. A. Balkema.*

Sheorey, P. R., Das, M. N., Brodia, S. K. and Singh, B. (1986). Pillar strength approaches based on a new failure criterion for coal seams. *International Journal of Mining and Geological Engineering,* 4: 273–290.

Simmons, J. (2003). New rock mechanics models needed for slope stability in open pit coal mines. *International Conference on Groundwater in Geological Engineering, Bled, Slovenia,* 22–26 September.

Simmons, J. (2013). Geotechnical support for open pit coal mining. *Australian Geomechanics,* 48(1): 85–100.

Simmons, J. V. (1995). Slope stability for open pit coal mining. Technical paper 9506, *Australian Geomechanics,* December, 45–57.

Simmons, J. V. and Simpson, P.J. (2007). Extension, stress and composite failure in bedded rock masses. *Proc. International Symposium on Rock Slope Stability in Open Pit Mining and Civil Engineering, Australian Centre for Geomechanics,* 213–223.

Sirovision™ User Manual, CAE Mining. (2013). [http://www.sirovision.com].

Terzaghi, R. D. (1965). Source of error in joint surveys. *Geotechnique,* 15: 287–304.

Townsend, J. M., Jennings, W. C., Haycocks, C., Neall, G. M., & Johnson, L. P. (1977). A relationship between the ultimate compressive strength of cubes and cylinders for coal specimens. *Proc. 18th U.S. Symposium on Rock Mechanics (eds. F. Wang and G. B. Clark), 4A6-1–4A6-6, Golden: Colorado School of Mines Press.*

Trueman, R. and Medhurst, T. P. (1994). The influence of scale effects on the strength and deformability of coal. *In M. Van Sint Jan (ed.). IV CSMR/Integral Approach to Applied Rock Mechanics, Santiago: Sociedad Chilena de Geotechnica,* 1: 103–114.

Vandergrift, T. and Garcia, J. (2005). Highwall mining in a multiple-seam, Western United States setting design and performance. *Proc. 24th International Conference on Ground Control in Mining, August 2–4, Morgantown, West Virginia, USA,* 218–224.

Venkateshwarlu, V., Ghose, A. K. and Raju, N. M. (1989). Rock mass classification for design of roof supports – A statistical evolution of parameters. *Mining Science and Technology*, 8: 97–106.

Verma, C. P., Thote, N. R. and Porathur, J. L. (2014). Highwall mining technology – a review of current status and design methods. *CIM Journal*, 5(4): 227–236.

Wagner, H. (1974). Determination of the complete load deformation characteristics of coal pillars. *Proc. 3rd International Congress on Rock Mechanics, International Society for Rock Mechanics (ISRM), Denver*, 1076–1082.

Wagner, H. and Galvin, J. M. (1979). Use of hydraulically placed PFA to improve stability in board and pillar workings in South African collieries. *Proc. of the Symposium on the Utilization of Pulverized Fuel Ash, Report No. CONF-7906215, Pretoria, South Africa.*

Wang, H., Poulsen, B. A., Shen, B., Xue, S. and Jiang, Y. (2011). The influence of roadway backfill on the coal pillar strength by numerical investigation. *International Journal of Rock Mechanics and Mining Sciences*, 48: 443–450.

Wang, L. G., Yamashita, S., Sugimoto, F., Pan, C. and Tan, G. (2003). A methodology for predicting the *in situ* size and shape distribution of rock blocks. *Rock Mechanics and Rock Engineering*, 36(2): 121–142.

Wilson, A. H. (1973). A hypothesis concerning pillar stability. *The Mining Engineer (London)*, June, 131(141): 409–417.

Xiao, G., Irvin, R. and Farmer, I. (1991). Water inflows into longwall workings in the proximity of aquifer rocks. *International Journal of Rock Mechanics and Mining Sciences and Geomechanics Abstracts*, 151: 9–13.

Zipf, R. K. (1999). Catastrophic collapse of highwall web pillars and preventative design measures. *Proc. 18th International Conference on Ground Control in Mining, Morgantown, West Virginia, USA*, 3–5 August, 18–28.

Zipf, R. K. (2005). Ground control design for highwall mining. *Proc. SME Annual Meeting, Salt Lake City, Utah, SME Preprint 05–82. Littleton, CO: Society for Mining, Metallurgy and Exploration Inc.*

Zipf, R. K. and Bhatt, S. K. (2004). Analysis of practical ground control issues in highwall mining. *Proc. 23rd International Conference on Ground Control in Mining, Morgantown, West Virginia, USA*, 3–5 August, 210–219.

Bibliography

Australian Mining Consultants. (1999). Report on gateroad drivage at Kenmare 2 for Roche Highwall Mining Pty Ltd. *Australian Mining Consultants Report No. AMC 399042.*

Barton, N. and Bieniawski, Z. T. (2008). RMR and Q – setting records straight. *Tunnels and Tunnelling International*, February, 26–29.

Hatherly, P., Medhurst, T. and MacGregor, S. M. (2008). Geophysical strata rating. ACARP End of Grant Report, C15019, Brisbane: *Australian Coal Association Research Programme (ACARP)*.

Itasca. (2006). *FLAC3D* – Fast Lagrangian analysis of continua in three-dimensions, Version 1.1, User's manual. *Itasca Consulting Group Inc., Minneapolis, Minnesota, USA.*

Jhanwar, Jagdish Chandra, Porathur, John Loui and Pal Roy, Pijush. (2014). Highwall mining in India – Part 3: Slope stability planning and management. *Journal of Mines, Metals and Fuels, IMME 2014, No. September–October, India*, 263–266.

Lorig, L. and Varona, P. (2000). Practical slope-stability analysis using finite-difference codes. *Slope stability in surface mining. Littleton, Colorado: SME*, Chapter 12, 115–124.

Medhurst, T., Hatherly, P. and Zhou, B. (2010). 3D geotechnical models for coal and clastic rocks based on the GSR. *10th Underground Coal Operators' Conference, University of Wollongong & the Australasian Institute of Mining and Metallurgy*, 40–49.

Palmstrom, A., Blindheim, O. T. and Broch, E. (2002). The Q-system – possibilities and limitations (in Norwegian). *Norwegian National Conference on Tunnelling, Norwegian Tunnelling Association*, 41.1–41.43.

Paul, A., Sing, A. K., Kumar, N. and Rao, D. G. (2009). Empirical approach for estimation of rock load in development workings of room and pillar mining. *Journal of Scientific and Industrial Research*, March, 68: 214–216.

Pit Slope Manual. 1977. *CANMET, Ottawa, Canada, (Chapter-Monitoring).*

Porathur, John L., Pal Roy, P., Verma, C. P. and Karekal, S. (2013). Extraction design for multiple seams highwall mining in India – A case example. *Proc. 47th US Rock Mechanics Symposium, San Francisco, USA*, 23–26 June, 3: 1656–1660.

Porathur, John Loui, Pal Roy, Pijush, Prakash, Amar and Jhanwar, Jagdish Chandra. (2013). Extraction design of locked-up coal by highwall mining in India. *23rd World Mining Congress and Expo 2013, Palais des congrès de Montréal, Canada*, 11–15 August.

Porathur, John Loui, Verma, Chandrani P., Pal Roy, P. and Sinha, A. (2013). Design and development norms for highwall mining in India. *Proc. 25th National Convention of Mining Engineers & the National Seminar on Policies, Statutes & Legislation in Mines – Recent Reforms and their Impacts on Indian Mining Industry, India, POSTALE.*

Porathur, John Loui, Srikrishnan, S., Verma, Chandrani Prasad, Jhanwar, J. C. and Pal Roy, P. (2014). Slope stability assessment approach for multiple seams highwall mining extractions. *International Journal of Rock Mechanics and Mining Sciences*, September, 70: 444–449.

Porathur, John Loui, Verma, Chandrani P. and Pal Roy, Pijush. (2014). Highwall mining in India – Part 1: Design methodology and review of performance, *Journal of Mines, Metals and Fuels, IMME 2014, No. September–October, India*, 245–253.

Prakash, Amar, Porathur, John Loui and Pal Roy, Pijush (2014). Highwall mining in India – Part 2: Subsidence management mechanism at mine level. *Journal of Mines, Metals and Fuels, IMME 2014, No. September–October, India*, 254–262.

Sowers, G. B. and Sowers, G. F. (1970). Introductory soil mechanics and foundations. *3rd ed. MacMillan, New York.*

Verma, C. P., Porathur, J. L., Thote, N. R., Pal Roy, P. and Karekal, S. (2014). Empirical approaches for design of web pillars in highwall mining: Review and analysis. *Geotechnical and Geological Engineering*, 32: 587–599.

Wilson, A. H. (1982). Pillar stability in longwall mining. *State-of-the-Art of Ground Control in Longwall Mining and Mining Subsidence* (ed. Y. P. Chugh and M. Karmis), *SME, New York*, 85–95.

Index

Milton Keynes UK
Ingram Content Group UK Ltd.
UKHW050448071024
449327UK00014B/287